SCIENCE AND PRACTICE OF STRENGTH TRAINING

Vladimir M. Zatsiorsky, PhD
The Pennsylvania State University

Human Kinetics

Library of Congress Cataloging-in-Publication Data

Zatsiorsky, Vladimir M., 1932–
Science and practice of strength training / Vladimir M. Zatsiorsky.
p. cm.
Includes bibliographical references and index.
ISBN 0-87322-474-4
1. Physical education and training. 2. Muscle strength. 3. Biomechanics. I. Title.
GV711.5.Z38 1995
613.7'11—dc20 94-40135
CIP

ISBN: 0-87322-474-4

Acquisitions Editor: Rik Washburn
Developmental Editor: Anne Mischakoff Heiles
Assistant Editors: Ed Giles, Kirby Mittelmeier
Copyeditor: Joyce Sexton
Proofreader: Amy Wilson
Indexer: Diana Witt
Typesetter and Layout Artist: Impressions, a Division of Edwards Brothers, Inc.
Text Designer: Judy Henderson
Cover Designer: Jody Boles
Photographer (interior): Yuri D. Sokoloff
Illustrators: Paul Toe, Gretchen Walters
Printer: Edwards Brothers, Inc.

Printed in the United States of America

10 9 8 7 6 5 4 3 2 1

Human Kinetics
P.O. Box 5076, Champaign, IL 61825-5076
1-800-747-4457

Canada: Human Kinetics, Box 24040, Windsor, ON N8Y 4Y9
1-800-465-7301 (in Canada only)

Europe: Human Kinetics, P.O. Box IW14, Leeds LS16 6TR, England
(44) 532 781708

Australia: Human Kinetics, 2 Ingrid Street, Clapham 5062, South Australia
(08) 371 3755

New Zealand: Human Kinetics, P.O. Box 105-231, Auckland 1
(09) 309 2259

To Rita, Betty, Michael, Stacia, Anastasia, and James
with thanks for your
love, courage, and support.
God bless you.

Contents

Foreword

During the past 45 years, strength training has advanced as a result of a more formal approach to its practice. A greater scientific understanding of the factors involved in the development of specific strength training programs has allowed more effective programs to be used by individuals interested in fitness as well as elite athletes training for sport competition. One fact that has been demonstrated over and over again is that weight training programs are specific to the goals and objectives of an individual. While science can provide a general direction to take with program development, it is still up to the individual or coach to fine-tune the program to meet the individual's needs. Thus the art of making informed decisions in this process is still a vital part of program development in strength training.

In this book, *Science and Practice of Strength Training*, my colleague and friend here at Penn State University, Dr. Vladimir Zatsiorsky, provides a unique perspective of basic and applied knowledge of strength training. Backed by his many years of direct practical and laboratory experience and work with weight lifting, weight training, and biomechanics in the former Soviet Union, as well as his work here in the United States, Dr. Zatsiorsky helps define, illuminate, and demonstrate theories of strength development that will be helpful to anyone interested in strength fitness. I feel that this textbook makes a valid contribution to the field of strength training and is a must for a personal library of books in this field of study. Dr. Zatsiorsky is an exceptional international scholar in the science of strength and conditioning, as evidenced by this book. Enjoy this text, as I have, for its insight and perspective on strength training.

William J. Kraemer, PhD
Penn State University
November 1994

Preface

This book is for readers who are interested in muscular strength and ways to enhance its development, that is, for

- coaches,
- students who plan to become coaches, and
- athletes who want to be self-coaches.

It developed from

- the documented experiences of more than 1,000 elite athletes—Olympic, world, continental, and national champions and record holders;
- Eastern European concepts of strength conditioning;
- international scientific data, and
- my experience as a consultant for the soviet national teams during seven Olympic cycles.

Science and Practice of Strength Training is designed for serious readers who are willing not only to remember and repeat but also to understand and put information to use.

On more than one occasion a coach or athlete has asked me for the best exercise, method, or training program to develop strength. I have no answer for such a question. Exercises (methods and programs) that are the best for all athletes, at all times, simply do not exist. They may only be the best for specific athletes under specific circumstances.

A sports training textbook is not like a cookbook where you can find a perfect recipe on any page. A complete understanding of the underlying factors and mechanisms of athletic training is of primary importance.

Two friends read an earlier draft of this book, an experienced American coach and a prominent American scientist. "It is interesting," they said politely. After that, their opinions were completely different. "Too much science," said the coach. "We do not use this language. The material is very difficult for an ordinary coach." "Too little science," said the scientist. "You do not prove many of your statements and recommendations" (this is not quite true; in many cases the proof is solid enough but is not available in English). "You are biased. You stick to one hypothesis or concept where several exist. You believe and promote the experience of some elite athletes, but you do not mention that other athletes have used other approaches."

I agree with the coach. However, this is a textbook and is meant to be studied, not simply read. I have attempted to write it as simply and clearly as possible. The complexity of the textbook is dictated by the complexity of its objective. Could you imagine a simple textbook in physics for future physicists? In sports, we are trying to improve the most wonderful creature of nature—a human being. Why should we be surprised about the complexity of the problem?

I also agree with the scientist. Currently, sport training is as much an art as it is a science. Practical experience is the greatest source of knowledge in the field; however, practical experience is always limited. New approaches will be available tomorrow and new experiences will begin. Current coaches should have the existing knowledge and should know how to combine scientific data to solve practical problems. At a minimum, they should know the most successful methods and the basic science. Other details and controversies will be described in later books.

This book is biased. It is heavily influenced by the Eastern European experience—predominantly in the former Soviet Union, former East Germany (German Democratic Republic), and Bulgaria—in the preparation of elite athletes. I do not underestimate the North American techniques, which are excellent. Thousands of wonderful American coaches and athletes, both amateur and professional, have proven this. Thus, reader, if you are one of these coaches or athletes, I am not going to teach you how to train. You have your own system—please be very careful about changing it. However, sport is based on the pursuit of excellence. If you want to be the best in the world, you need to avoid mistakes and maybe even invent something new: exercises, methods, or a training system. Knowledge gained by other coaches and athletes from other countries will help you.

This text is intended to be comprehensive. However, one important issue is not addressed—drug abuse. Unfortunately, in order to become bigger and stronger, many athletes, including bodybuilders and teenagers caught up in the bodybuilding craze, use drugs, especially black-market anabolic steroids. This practice is harmful to their health, unethical (in sport), and illegal. The use of drugs, steroids included, is banned by the International Olympic Committee and international amateur athletic federations. The distribution of steroids for nonmedical purposes is forbidden by law in many countries, including the United States and Canada. This book is written to show you how to train without drugs. It is important to realize that the vast majority of elite athletes have never used drugs.

The facts I present here were observed in soviet athletes. With the collapse of the U.S.S.R., training practices did not change, and athletes from other former soviet republics trained similarly to their Russian counterparts. There is no easy way to refer to these athletes now by country or nationality. As you read this book, you will notice that often they are described as soviet/Russian. I mean to refer to those athletes of the various nationalities that used to comprise the U.S.S.R.

Interesting special material is designated throughout the book by the symbol ■. A short index of recommended reading—books and review papers only, rather than an extensive reference list—is provided in this book. A comprehensive list of related publications, numbering more than a thousand, would be too long for this book, which is for coaches and not scientists. In addition, substantial portions of the literature are not available in English or hardly available at all; much of this literature has been published in small quantities for restricted distribution among elite coaches working with national Olympic teams.

Prof. Vladimir M. Zatsiorsky, PhD

Acknowledgments

Numerous people helped me in important ways in preparing the manuscript for this book.

I am most grateful to Dr. Richard C. Nelson, head of the Biomechanics Lab. This book would not have been written without his invaluable support and help. Thank you, Dick. Special thanks goes to Dr. Robert J. Gregor (University of California, Los Angeles) and Dr. Benno M. Nigg (University of Calgary), for inviting me as a visiting researcher at their laboratories. The book was written in part during this time.

I thank Dr. William J. Kraemer for his useful comments. Dr. J.M. Kots helped me to avoid inaccuracies in describing the electrostimulation method of strength training, a method he invented. Dr. V.B. Issurin at the Wingate Institute of Physical Culture, Israel, taught me much about training periodization. The taxonomy of mesocycles accepted in this book is due to him.

I am especially indebted to the Human Kinetics external reviewers, to Peter M. McGinnis, and to the second reviewer, unknown to me, for correcting errors in the manuscript and improving its readability.

Sherry Werner, Cathy Lendrim, Clare Johnson, and Jeff Bauer read the manuscript chapters and transformed my Russian version of the English language into acceptable English. A special thanks is extended to Peter Brown, whose help as a grammarian and stylist is impossible to overestimate. His intense efforts shine throughout the entire book.

To all who helped, *spasibo* (thank you).

I do not thank my wife, however, because to thank her would be the same as thanking myself.

Symbols and Abbreviations

BW	Body weight
CF_{mm}	Maximum competition weight
DF_m	Gain in maximal force
EMG	Electromyography
EMS	Electrical stimulation of muscles
ESD	Explosive strength deficit
F	Force
F_m	Maximal force attained when the magnitude of a motor task parameter is fixed
F_{mm}	Maximum maximorum force attained when the magnitude of a motor task parameter is altered
FT	Fast-twitch muscle fibers
g	Acceleration due to gravity
IAP	Intraabdominal pressure
IES	Index of explosive strength
LBPS	Low-back pain syndrome
MSD	Muscle strength deficit
MU	Motor unit
N	Newton; the unit of force
P_m	Maximal performance attained when the magnitude of a motor task parameter is fixed
P_{mm}	Maximum maximorum performance attained when the magnitude of a motor task parameter is altered
RC	Reactivity coefficient
RFD	Rate of force development
RM	Repetition maximum

ST Slow-twitch muscle fibers

T_m Time to peak performance

TF_{mm} Maximum training weight

V_m Maximal velocity attained when the magnitude of a motor task parameter is fixed

V_{mm} Maximum maximorum velocity attained when the magnitude of a motor task parameter is altered

PART I

BASIS OF STRENGTH CONDITIONING

The primary goal of this book is to provide readers with practical recommendations, or a *prescription,* for training athletes. Practical advice, however, cannot be given without first providing *descriptions* of what should be trained and why some methods are better than others. Part 1 of the book describes theory, while Part 2 covers methods of strength training.

The first part, which is entirely descriptive, develops several concepts in a natural, sequential order. Chapter 1 is introductory and provides an overview of the principles of training theory: It describes the peculiarities of adaptation to a physical load; discusses two prevailing theories of training—the supercompensation theory and the fitness-fatigue theory—both of which are widely and enthusiastically embraced as effective methods in spite of their simplicity; and spells out the nomenclature of training effects. Although the concepts and terminology introduced in this chapter are used throughout the book, the chapter is self-contained and presumes that the reader has no prior scientific knowledge.

Chapters 2 and 3 address the factors that determine muscle strength. It is assumed that readers have some knowledge of exercise physiology and sport biomechanics, or at least are acquainted with basic muscle physiology. Readers who are not familiar with this material, however, should not be discouraged from reading the book; the main concepts are explained in a format intelligible for a reader with a minimal background in exercise and sport science. Readers who do have trouble understanding chapters 2 and 3 need not read them in one sitting, but can return to them later while reading the balance of the book.

Chapter 2 lays the foundation for the very notion of muscular strength, classifying and explaining the evidence collected by measuring muscle force. It introduces the concept of maximal muscular performance, as well as two main relationships—parametric and nonparametric—and defines the notion of muscular strength. It then follows with a detailed discussion of various factors involved in motor tasks, such as resistance, the time available for force development, movement velocity, movement direction, and body posture. The integrating idea for these diverse issues is rather simple and straightforward: exercise specificity. For training to be effective, the training exercises should be similar to the main sport activity, and the exercise similarity should be established according to the criteria discussed in this chapter.

Chapter 3 addresses the issue of muscle strength from another standpoint: that of the performer rather than the motor task. Some people possess greater strength than others. Why? What properties do elite athletes have that allow them to be exceptional? The internal factors determining muscle strength are latent. Hence, they can be identified only by using a physiological approach. If we are able to identify them, we open the road to goal-directed training of these primary factors, so the exercises and methods addressed here will center on specific targets rather than on strength in general. This chapter is based mainly on facts and theories originated by exercise physiologists. Two main groups of internal factors are discussed: muscular and neural. Among the muscular factors, primary attention is given to the muscle dimension and its counterpart, body weight. Other factors, including nutrition and hormonal status, are briefly highlighted, too. The neural mechanisms, such as intra- and intermuscular coordination, are reviewed in the later sections. Chapter 3 is essential for understanding training methods.

CHAPTER 1

Basic Concepts of Training Theory

Strength conditioning theory is part of a broader field of knowledge, the science of training athletes, also termed training science or theory of sport training. Training science courses cover the principal components of athlete preparation, including conditioning (not only for strength but also for speed, endurance, flexibility, and other motor abilities), sport technique learning, and periodization. Throughout this book, the concepts and approaches developed within the framework of training science are extensively utilized. This chapter introduces you to the issues of training in general. The ideas and terminology you encounter here will be used in the remainder of the book.

Adaptation as a Main Law of Training

If a training routine is planned and executed correctly, the result of systematic exercise is improvement of the athlete's physical fitness, particularly strength, as the body adapts to physical load. In a broad sense, the word *adaptation* means the adjustment of an organism to its environment. If the environment changes, the organism changes to better survive in these new conditions. In biology, adaptation is considered one of the main features of living species.

Exercise or regular physical work is a very powerful stimulus for adaptation. The major objective in training is to induce specific adaptations in order to improve sport performance results. This requires adherence to a carefully planned and executed training program. From the practical point of view, the following four features of the adaptation process assume primary importance for sport training:

1. The stimulus magnitude (overload)
2. Accommodation
3. Specificity
4. Individualization

Overload

To bring about positive changes in an athlete's state, an exercise *overload* must be applied. The training adaptation takes place only if the magnitude of the training load is above the habitual level. During the training process, there are two ways to induce the adaptation. One is to increase the training load (intensity, volume) while continuing to employ the same drill, for example, endurance running. The other is to change the drill, provided that the exercise is new and the athlete is not accustomed to it.

If an athlete uses a standard exercise with the same training load over a very long time, there will be no additional adaptations and the level of physical fitness will not substantially change (Figure 1.1). If the training load is too low, detraining occurs. In elite athletes, many training improvements are lost within several weeks, even days, if an athlete stops exercising. During the competition period, elite athletes cannot afford complete passive rest for more than 3 days in a row (typically only 1 or 2 days).

Training loads can be roughly classified according to their magnitude as

- *stimulating*—the magnitude of the training load is above the neutral level and positive adaptation may take place;
- *retaining*—the magnitude is in the neutral zone at which the level of fitness is maintained; and
- *detraining*—the magnitude of the load leads to a decrease in performance results, in the functional capabilities of the athlete, or both.

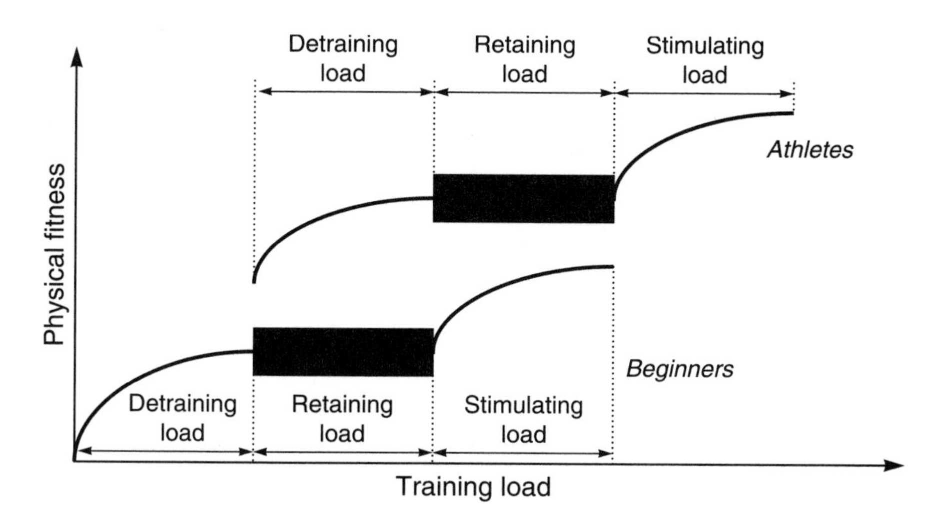

Figure 1.1 Dependence between training load (detraining, retaining, stimulating) and level of physical fitness. Rectangles indicate the neutral zones (retaining load) corresponding to small fluctuations in the training load at which the level of fitness is basically not changed. Note the "stepladder" effect showing a change in the adaptation curve with a change in the training stimulus. A training load that leads to the detraining of qualified athletes may be extremely high for beginners.

■ *Overload*

Identical triplets possessed equal levels of strength; each was able to lift a 57.5-kg barbell one time. They began to exercise with a 50-kg barbell, lifting the barbell in one set until failure five times. After a period of time, the athletes adapted to the training routine, their preparedness improved, and they were able to lift a 60-kg barbell one time. However, despite continued training, they did not make further performance gains because they accommodated to the training program.

At this stage, the three athletes made different decisions. Athlete A decided to increase the training load (weight lifted, the number of repetitions in a set, the number of sets) or change the exercise. The new load was a stimulating load for this athlete and the performance improved. Athlete B continued to employ the previous routine and the performance results were unchanged (retaining load). Athlete C decreased the training load and this athlete's strength performance declined (detraining load).

Because the preparation of elite athletes requires 8 to 12 years, the need for a constant increase in training loads, considered necessary for positive adaptation, leads to extremely demanding training programs. The training load of elite athletes is roughly 10 times greater than that of beginners having 6 months of training experience. For instance, the year-round training mileage of elite cross-country skiers is between 8,000 and 12,000 km. For beginners, it is about 1,000 km. Elite weight lifters (Bulgarian) lift around 5,000 tons a year; the load for novices is only 1/10th or 1/12th of this level.

Accommodation

If athletes employ the same exercise with the same training load over a long period of time, performance gains decrease (see Figure 1.2). This is a manifestation of the biological law of accommodation, often considered a general law of biology. According to this law, the response of a biological object to a given constant stimulus decreases over time. By definition, *accommodation* is the decrease in response of a biological object to a continued stimulus. In training, the stimulus is physical exercise.

Because of accommodation, it is inefficient to use standard exercises or a standard training load over a long period of time. Training programs must vary. At the same time, because of the specificity of training adaptations, the employed exercises should be as close as possible to the main

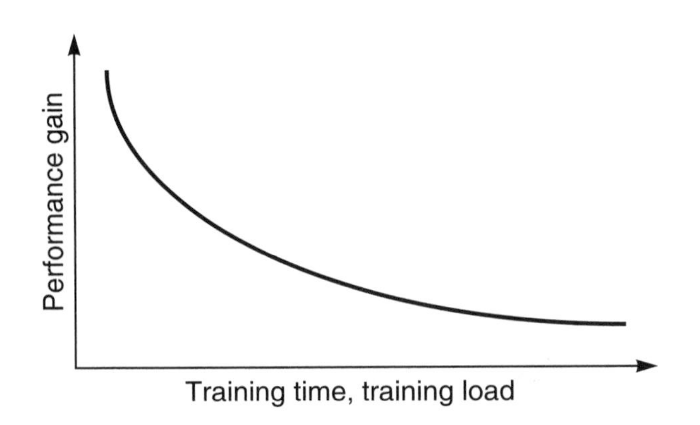

Figure 1.2 Dependence of performance gain on time of training or training load. As a result of accommodation, the gain decreases.

sport exercise in muscular coordination and physiological demand. The highest transfer of training result occurs with the use of specific exercises. These two requirements lead to one of the main conflicts in training elite athletes: Training programs should be both variable, to avoid accommodation, and stable, to satisfy the demand for specificity.

To avoid or decrease the negative influence of accommodation, training programs are periodically modified. In principle, there are two ways to modify training programs:

- Quantitative—changing training loads (for instance, the total amount of weight lifted) and
- Qualitative—replacing the exercises.

Qualitative changes are very broadly used in the training of elite athletes, at least by the most creative.

Specificity

Training adaptations are highly specific. It is well known that strength training increases both muscle mass and strength while endurance running induces positive changes such as increases in aerobic capacity. Because of adaptation specificity, the exercises and training in various sports are different.

Specificity may be described in another way, as an issue of transfer of training. Imagine, for example, a group of young athletes who have trained over a certain period of time with one exercise, exercise A, squatting with a barbell. Finally, their performances improve. Let's suppose that the gain is the same for all the athletes, say 20 kg. What will happen with the performances of these athletes in other exercises, such as the standing vertical jump, sprint dash, or freestyle swimming (exercises B, C, and D)? We may predict that the results in these exercises will improve to different degrees. The gain may be substantial in the standing jump, relatively small in sprint running, and nearly zero in swimming. In other words, the transfer of training results from exercise A to exercises B, C, and D is variable.

■ *Transfer of Training Results: Why Is It Important?*

The first books about athlete preparation, published in the last century, make interesting reading. The preparation for competition consisted of the main sport exercise and nothing else. If one competed in the 1-mile run, workouts consisted of only 1-mile runs. This was called "training."

However, coaches and athletes soon understood that such preparation was not sufficient. To run a mile successfully, an athlete must not only have stamina but must also possess appropriate sprinting abilities, good running technique, and strong and flexible muscles and joints. It is impossible to develop these abilities by running the same fixed distance repeatedly. As a consequence of this realization, training strategies were changed. Instead of multiple repetitions of a single exercise, many auxiliary exercises were adopted into training programs to improve the abilities specific to a given sport. The general concept of training changed.

The question then arises: How do you choose more efficient exercises that result in a greater transfer of training effect from the auxiliary to the main sport movement? Consider the following problems:

1. Is long-distance running a useful exercise for endurance swimmers? For cross-country skiers? For race walkers? For bicyclists? For wrestlers?
2. To improve the velocity of fast pitches, a coach recommends that pitchers drill with baseballs of varying weight, including heavy ones. What is the optimal weight of the ball for training?
3. A conditioning coach planning a preseason training routine for wide receivers must recommend a set of exercises for leg strength development. The coach may choose between several groups of exercises or combine exercises from different groups. The exercise groups are

   - one-joint isokinetic movements, such as knee extension and flexion, on exercise apparatus,
   - similar one-joint drills with free weights,
   - barbell squats,
   - isometric leg extension,
   - vertical jumps with additional weights (heavy waist belts),
   - uphill running, and
   - running with parachutes.

   Which exercise is more effective? In other words, when are the transfer of training results greater?

The transfer of training gains can differ greatly even in very similar exercises. In an experiment, two groups of athletes performed an isometric knee extension at different joint angles, 70° and 130° (a complete leg extension corresponds to 180°). The maximal force values, F_m, as well as the force gains, ΔF_m, observed at different joint angles were varied (Figure 1.3, a and b).

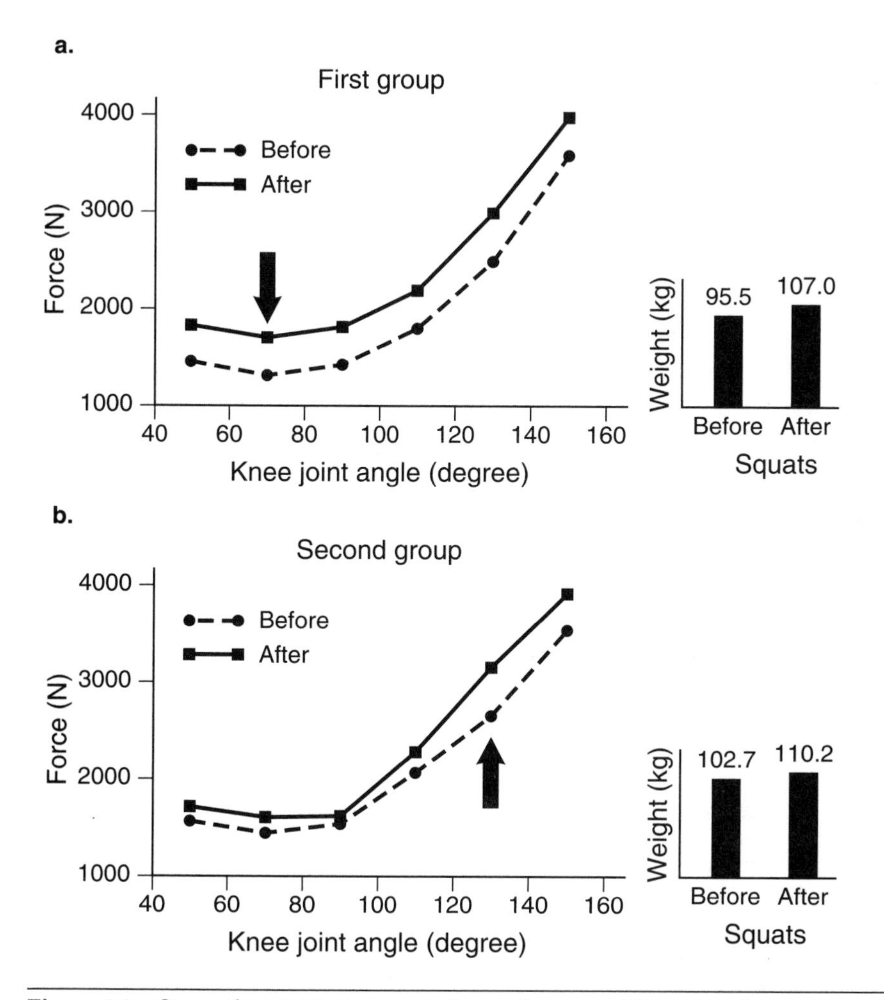

Figure 1.3 Strength gains in two experimental groups. The vertical arrows show the angles at which isometric training took place. Strength was measured in leg extension as well as in barbell squats. Note. The data are from "Transfer of Cumulative Training Effects in Strength Exercises" by V.M. Zatsiorsky and L.M. Raitsin, 1974, *Theory and Practice of Physical Culture*, (6), pp. 7-14. Reprinted by permission from the journal.

The strength gains at various joint positions were different for the two groups. For the subjects in the first group, who exercised at the 70° knee-joint angle (see Figure 1.3a), the strength gains in all joint positions were almost equal. The transfer of training results from the trained body posture (70°) to untrained positions (other joint angles) was high. In the athletes of the second group, who trained at the 130° knee-joint angle (see Figure 1.3b), transfer of training gains was limited to the neighboring joint angles: The strength gain was low for small joint angles (compare strength gains in angles 130° and 90°). The same held true for barbell squats. In the first group, the strength gain in the trained body posture was 410 ± 170 N and in squatting it was 11.5 ± 5.4 kg. In the second group, the strength in the trained posture increased by 560 ± 230 N; however, in spite of such a high gain, the barbell squat performance improved by only 7.5 ± 4.7 kg. The strength gain in the trained posture in the second group was higher (560 ± 230 N vs. 410 ± 170 N), but the improvement in the barbell squats was lower (7.5 ± 4.7 kg vs. 11.5 ± 5.4 kg) due to minimal transfer of training results.

As performances in different exercises have different modalities (force, time, distance) and are not directly comparable, a dimensionless unit should be employed to estimate the *transfer of training result*. Such a unit is a result gain expressed in standard deviations:

$$\text{Result gain} = \frac{\text{Gain of performance}}{\text{Standard deviation of performance}}$$

For instance, if the average performance of a group is 60 ± 10 kg (average ± standard deviation) and the performance of an athlete is improved as a result of training by 15 kg, the athlete's personal gain equals 15/10 or 1.5 standard deviation. For the estimation of transfer, a ratio of the gains in nontrained exercises (exercises B, C, and D) and the trained exercise (exercise A) is employed. The coefficient of the transfer of training is, by definition, the ratio:

$$\text{Transfer} = \frac{\text{Result gain in nontrained exercise}}{\text{Result gain in trained exercise}}$$

Both gains are measured in standard deviations. The higher the ratio, the greater the transfer of training results. If the transfer is low, the effect of training is specific. In the example from Figure 1.3, training effects were more specific for the group that performed exercise at the 130° knee-joint angle.

Specificity of adaptation increases with the level of sport mastership. The higher an athlete's level of fitness, the more specific the adaptation. The transfer of training gain is lower in good athletes; for beginners, almost all exercises are useful. It is possible to improve the strength, speed, endurance, and flexibility in people with extremely low physical fitness through simple calisthenics. The performance of beginning bicyclists can be improved by squatting with a barbell. Elite athletes should use more specific exercises and training methods to increase competitive preparedness.

■ *Calculating the transfer of training results*

In the experiment, the following data were recorded (Figure 1.3):

Test	Before	After	Gain of performance	Result gain	Transfer
		Group 1 (Isometric training at an angle of 70°)			
Force at an angle 70°, N	1310 ± 340	1720 ± 270	410 ± 170	410/340 = 1.2	
Squatting, kg	95.5 ± 23	107 ± 21	11.5 ± 5.4	11.5/23 = 0.5	0.5/1.2 = 0.42
		Group 2 (Isometric training at an angle of 130°)			
Force at an angle 130°, N	2710 ± 618	3270 ± 642	560 ± 230	560/618 = 0.91	
Squatting, kg	102 ± 28	110 ± 23	7.5 ± 4.7	7.5/28 = 0.27	0.27/0.91 = 0.30

Note the results:

Characteristics	Superior group	Comparison
Gain of performance in trained exercise	Second	560 vs. 410 N
Result gain in trained exercise	First	1.2 vs. 0.91 SD
Transfer of training results	First	0.42 vs. 0.30
Gain of performance in nontrained exercise	First	11.5 ± 5.4 vs. 7.5 ± 4.7 kg

Because of the higher transfer of training results, the method used to train the first group better improved the squatting performance.

Individualization

All people are different. The same exercises or training methods elicit a greater or smaller effect in various athletes. Innumerable attempts to mimic the training routines of famous athletes have proven unsuccessful. Only the general ideas underlying noteworthy training programs, not the entire training protocol, should be understood and creatively employed. The same holds true for average values derived from training practices and scientific research. Coaches and athletes need to use an average training routine cautiously. Only average athletes, those who are far from excellent, prepare with average methods. A champion is not average, but exceptional.

Generalized Theories of Training

Generalized training theories are very simple models that coaches and experts use broadly to solve practical problems. These models include only the most essential features of sport training and omit numerous others. These generalized theories (models) serve as the most general concepts for coaching. Coaches and athletes use them especially for conditioning and also for planning training programs.

One-Factor Theory (Theory of Supercompensation)

In the one-factor theory, the immediate training effect of a workout is considered as a depletion of certain biochemical substances. The athlete's disposition toward a competition or training, called an athlete's *preparedness,* is assumed to vary in strict accordance with the amount of a substance available for an immediate use. There is evidence in exercise and sport science literature that certain substances are exhausted as a result of strenuous training workouts. The best known example is glycogen depletion after hard anaerobic exercise.

After the restoration period, the level of the given biochemical substance is believed to increase above the initial level. This is called *supercompensation,* and the time period when there is an enhanced level of the substance is termed the *supercompensation phase* (Figure 1.4).

If the rest intervals between workouts are too short, the level of an athlete's preparedness decreases (see Figure 1.5a). If the rest intervals between consecutive workouts are the right length, and if the next training session coincides in time with the supercompensation phase, the athlete's preparedness advances (Figure 1.5b). Finally, in the case of very long intervals between sessions, an athlete's physical abilities do not change (Figure 1.5c). A coach or athlete should avoid time intervals between serial training sessions that are either too short or too long, and instead seek

- optimal rest intervals between successive training sessions and
- an optimal training load in each workout.

The aim in selecting these intervals and loads is to ensure that a subsequent training session coincides with the supercompensation phase.

Within the framework of this theory, more sophisticated variations of the training schedule are also acceptable. One that is popular among

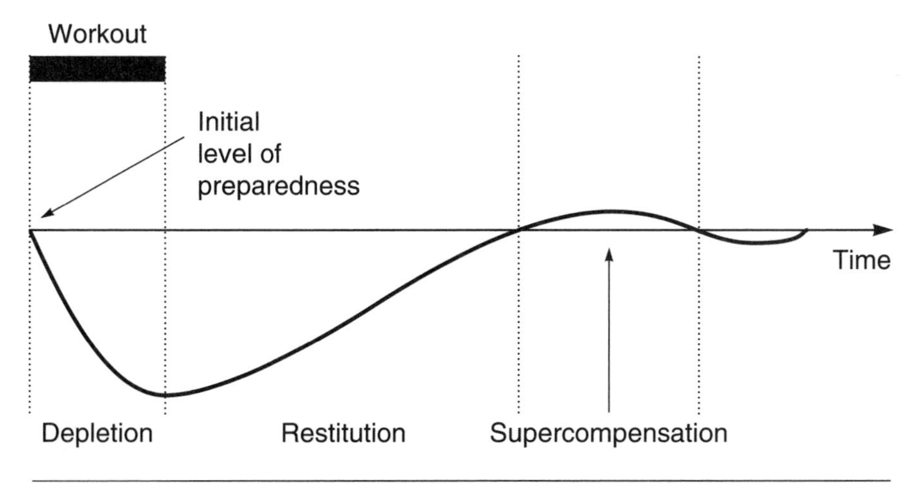

Figure 1.4 Time course of the restoration process and athlete's preparedness (AP) after a workout according to the supercompensation theory. The vertical axis is both for the amount of substance and for the level of preparedness. According to the model, the two curves coincide.

coaches, the overloading microcycle (or impact microcycle), is shown in Figure 1.6. In this case, after several training sessions with high training loads and short time intervals between sessions, a relatively long period of rest is included. The common belief is that such a training routine produces a final supercompensation that is greater than normal (compare Figures 1.5b and 1.6).

For several decades, the supercompensation model has been the most popular training theory. It has been described in many textbooks and is widely accepted by coaches. In spite of its popularity, it deserves critical scrutiny.

The very existence of the supercompensation phase for a majority of metabolic substances has never been experimentally proven. For some metabolites, like glycogen, after-exercise depletion has been definitely demonstrated. It is possible to induce glycogen supercompensation by combining a proper training routine with carbohydrate loading. This procedure, however, cannot be reproduced regularly and is used only before important competitions, not for training. The concentrations of other biochemical substrata whose role in muscular activity has been proven to be very important, for example, adenosine triphosphate (ATP), do not change substantially even after very hard exercise. The restoration of initial levels of different metabolic substances requires unequal amounts of time. It is absolutely unclear which criteria (substances) one should use for selecting proper time intervals between

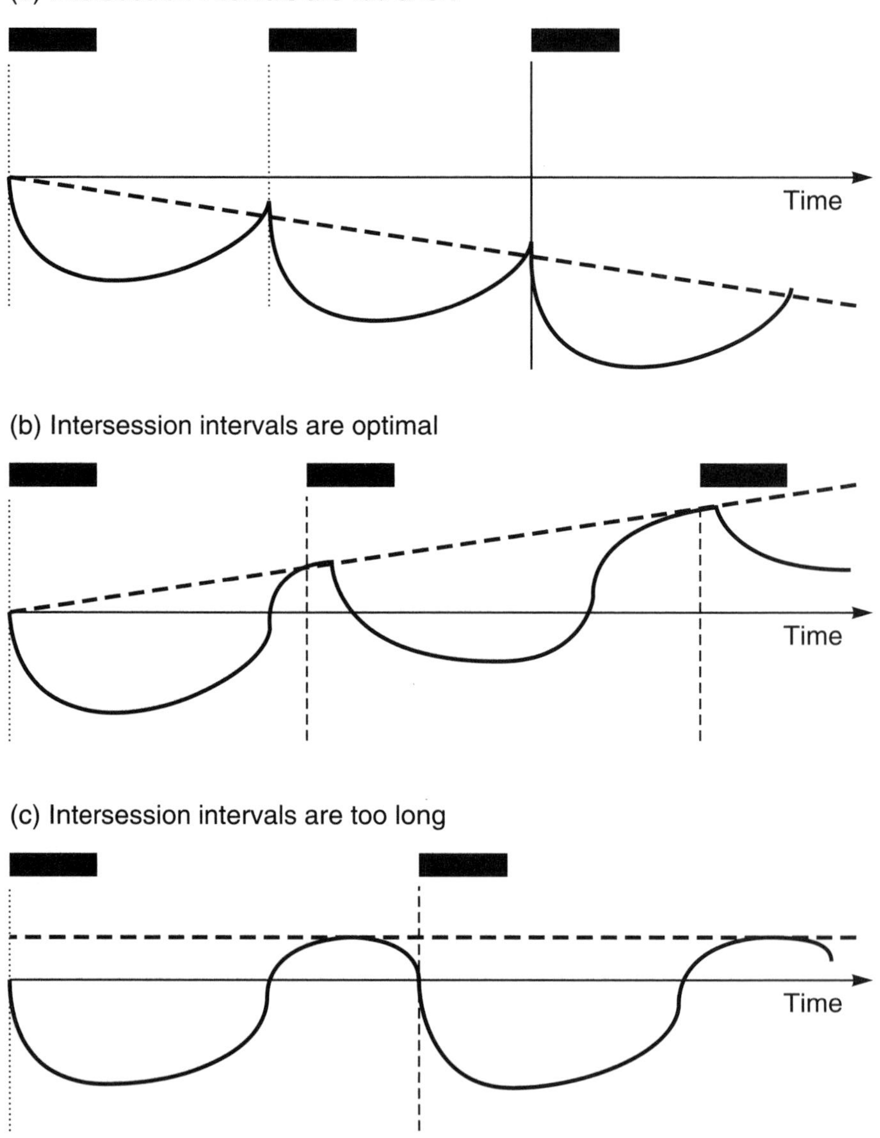

Figure 1.5 The supercompensation theory. The vertical axis is both for the amount of substance and for the level of preparedness. There are three main situations with rest intervals between sequential training workouts: (a) The intervals are too short and the level of athlete preparedness decreases due to accumulated fatigue; (b) The intervals are optimal and the ensuing workouts match with the supercompensation phase; and (c) The intervals are too long and there is no stable training effect.

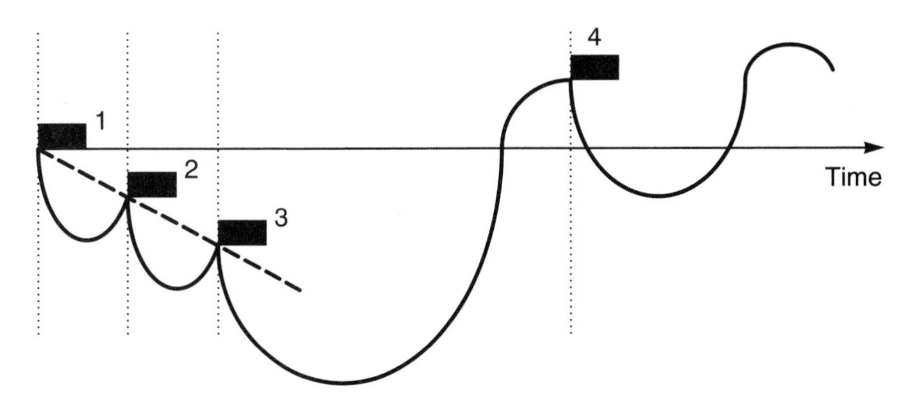

Figure 1.6 The "overloading microcycle" from the point of view of the supercompensation theory. Rest intervals between the first three training sessions are intentionally too short to allow full restoration, so fatigue is accumulated. The interval between the third and fourth training workouts is longer than usual but optimal for the situation. The next workout coincides with the supercompensation phase after the first three training sessions.

consecutive workouts. In general, the theory of supercompensation is too simple to be correct. Over the last few years it has lost much of its popularity.

Two-Factor Theory (Fitness-Fatigue Theory)

The two-factor theory of training is more sophisticated than the supercompensation theory. It is based on the idea that preparedness, characterized by the athlete's potential sport performance, is not stable but rather varies with time. There are two components of the athlete's preparedness: those that are slow-changing and those that are fast-changing. The term *physical fitness* is used for slow-changing motor components of the athlete's preparedness. Physical fitness does not vary substantially over several minutes, hours, or even days. However, as a result of fatigue, psychological overstress, or a sudden illness such as flu, an athlete's disposition toward competition may change quickly. An athlete's preparedness is sometimes thought of as a set of latent characteristics that exist at any time but can be measured only from time to time. According to the two-factor model, the immediate training effect after a workout is a combination of two processes:

1. Gain in fitness prompted by the workout and
2. Fatigue.

After one workout, an athlete's preparedness

- ameliorates due to fitness gain, but
- deteriorates because of fatigue.

The final outcome is determined by the summation of the positive and negative changes (Figure 1.7).

The fitness gain resulting from one training session is supposed to be moderate in magnitude but long lasting. The fatigue effect is greater in magnitude, but relatively short in duration. For most crude estimations, it is assumed that for one workout with an average training load, the durations of the fitness gain and the fatigue effect differ by a factor of three: The fatigue effect is three times shorter in duration. This implies that if the negative impact of fatigue lasts, for instance, 24 hr, the positive traces from this workout will remain through 72 hr.

The time course of the immediate training effect after a single workout can be described by the equation:

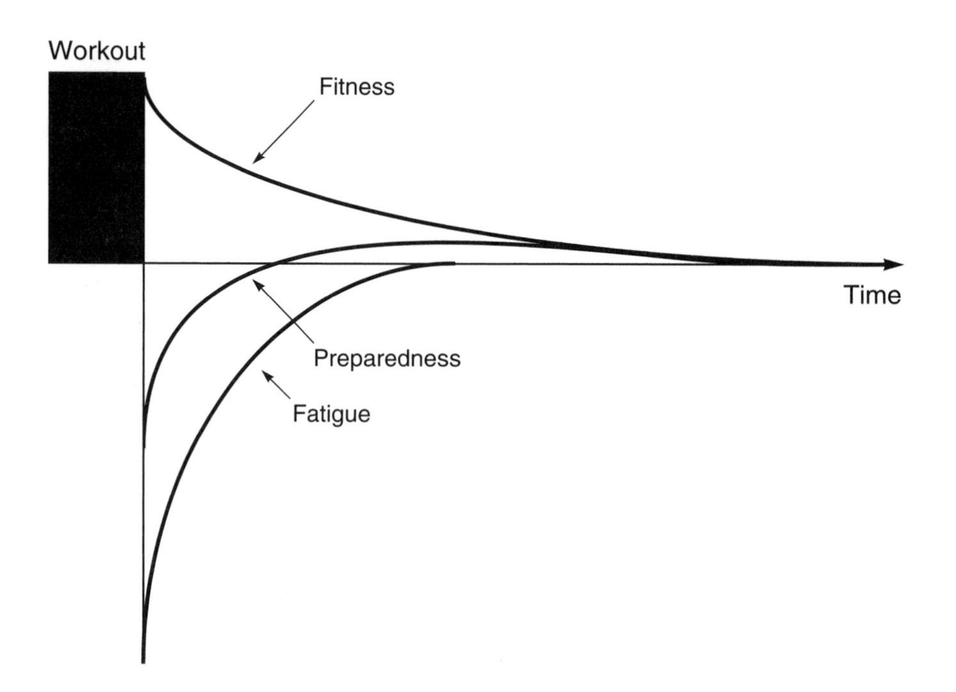

Figure 1.7 Two-factor theory (model) of training. The immediate training effect of a training session is characterized by the joint action of two processes, fitness gain and fatigue. Athlete preparedness improves because of fitness gain and worsens because of fatigue.

$$\text{Preparedness} = P + (F_1 \cdot e^{-k_1 t}) - (F_2 \cdot e^{-k_2 t})$$

where
P is the initial level of preparedness before the training workout;
F_1 is the fitness gain;
F_2 is the fatigue effect estimated immediately after the workout;
t is time;
k_1 and k_2 are time constraints; and
e is the base of the system of natural logarithms, approximately 2.718.

■ One- and Two-Factor Models of Training

These models help coaches to grasp and visualize the timing of workout-rest intervals during preparation of athletes and to view training as an organized process rather than a chaotic sequence of drill sessions and rest periods.

Imagine two coaches with different coaching philosophies. Coach A strictly adheres to the one-factor theory of training and is trying to schedule a training session for when (in his estimation) the supercompensation phase takes place. Coach B prefers the two-factor theory of training and is looking for rest intervals that are long enough for proper restoration and, at the same time, short enough to maintain the acquired physical fitness level. At times the training plans of the two coaches may look similar, but the underlying philosophies are not the same. You would see the greatest differences in plans for "peaking" periods that take place immediately before important competitions. Coach A would probably recommend that his athletes decrease the number of training sessions (but not the load during the sessions) in order to compete at the climax of the supercompensation phase. For instance, in accordance with the one-factor theory, he has the athletes train only two or three times during the final week before the main competition, with each workout containing a relatively large load. Coach B, on the other hand, prefers that her athletes maintain acquired preparedness, avoid fatigue, and participate in several warm-up type training sessions. The idea here is to decrease the training load during each session rather than the number of workouts.

According to the two-factor theory of training, the time intervals between consecutive training sessions should be selected so that all the negative traces of the preceding workout pass out of existence but the positive fitness gain persists. This model has become rather popular among coaches and is used predominantly to plan training, especially during the final training days before a competition.

Training Effects

Training effects, that is, changes that occur within the body as a result of training, can be further classified as follows:

- *Acute effects* are the changes that occur during exercise.
- *Immediate effects* are those that occur as a result of a single training session and that are manifested soon after the workout.
- *Cumulative effects* occur as a result of continued training sessions or even seasons of training.
- *Delayed effects* are those manifested over a given time interval after a performed training routine.
- *Partial effects* are changes produced by single training means (e.g., bench press exercise).
- *Residual effects* are defined as the retention of changes after the cessation of training beyond time periods during which adaptation can take place.

Summary

The major objective in training is to induce specific adaptations toward the improvement of athletic performance. In strength training, adaptation means the adjustment of an organism to exercise (physical load). If a training program is properly planned and executed, an athlete's strength improves as a result of adaptation.

Training adaptation takes place when the training load is above normal or the athlete is not accustomed to an exercise. Training loads are roughly classified as stimulating, retaining, and detraining loads. In order to induce the required adaptation:

1. An exercise overload must be applied.
2. The exercises and training protocol must be specific (corresponding to the main sport exercise).
3. Both exercises and training load (intensity, volume) should vary over time periods. When the same exercise with the same training load is employed over a long period of time, performance gains decrease. This is called accommodation.
4. Training programs must be adjusted individually to each athlete. Remember that all people are different.

To plan training programs, coaches use simple models that are based on only the most essential features. These models are known as generalized theories of training.

The theory of supercompensation, or one-factor theory, is based on the idea that certain biochemical substances are depleted as a result of training workouts. After the restoration period, the level of the substance increases above the initial level (supercompensation). If the next workout takes place during the supercompensation phase, the athlete's preparedness increases. In the fitness-fatigue theory (two-factor theory), the immediate effect after a workout is considered a combination of (a) fitness gain prompted by the workout and (b) fatigue. The summation of positive and negative changes determines the final outcome.

The effects of training can be classified as acute, immediate, cumulative, delayed, partial, or residual.

CHAPTER 2

Task-Specific Strength

If the goal is knowing *how* an athlete must train to achieve the best results, the steps along the way are to first know *what* it is that should be trained and to understand *why* the training must be performed in a prescribed way. To properly understand training, you must first clearly understand the notion of muscular strength per se.

In this chapter you will examine the definition of muscular strength and then learn the main factors that determine its development. When an athlete sincerely attempts a maximal effort, the resulting force depends on both the motor task and the athlete's abilities. Therefore, we will look at the determining factors as they compare across tasks and then, in chapter 3, examine the determining factors as they compare across athletes.

Elements of Strength

If an athlete were asked to produce a high force against a penny, the effort would fail. In spite of the best effort, the magnitude of force would be rather small. We may conclude that the magnitude of muscular force depends on the external resistance provided. Resistance is one of the factors that act to determine the force generated by an athlete, but only one. Other factors are also important, and here we explore them in detail.

Maximal Muscular Performance

Imagine an athlete who is asked to put a shot several times, making different efforts in various attempts. According to the laws of mechanics, the throwing distance is determined by the position of the projectile at release and its velocity (both magnitude and direction) at that moment. Let's suppose that the release position and release angle of the shot are not changed in different attempts. In this case, the throwing distance (performance) is determined only by the initial velocity of the projectile. As the subject throws the shot with different efforts in different attempts, the throwing distance is maximal in only one case. This is the individual's maximal muscular performance (maximal distance, maximal velocity). The symbol P_m (or V_m for maximal velocity, F_m for maximal force) will be used throughout this book to specify *maximal muscular performance.*

Parametric Relationships

At the next stage of the experiment, the athlete puts the shots with maximal effort, trying to achieve the best possible result. However, instead of putting the men's shot (7,257 g), the athlete puts the women's shot (4,000 g). The shot velocity is obviously greater using the lower weight. Two different values of V_m, one for the men's shot and one for the women's shot, are registered as a result of this experiment.

In science, a variable that determines the outcome of the experiment (such as mass or distance) or the specific form of a mathematical expression is termed a *parameter*. In other words, the parameter is an independent variable that is manipulated during the experiment. We may say that in the last example, the experimental parameter (shot mass) was changed. If the shot mass (parameter) is changed in a systematic way, for instance in the range from 0.5 kg to 20 kg, the maximal muscular performance (P_m, V_m, F_m) for each used shot will be different.

The dependent variables, in particular F_m and V_m, are interrelated. The relationship between V_m and F_m is called the *maximal parametric relationship*. The term *parametric* is used here to stress that V_m and F_m were changed because the values of the motor task parameter were altered. The parametric relationship between V_m and F_m is typically negative. In the throw of a heavy shot, the force applied to the object is greater and the velocity is less than in the throw of a light shot. The greater the force F_m, the lower the velocity V_m. The same holds true for other motor tasks (Figure 2.1, a and b)

■ *Parametric Relationships*

A coach suggested that athlete-cyclists change their gear ratios during training. The higher the ratio, the greater the force applied to the pedals and the lower the pedaling frequency. The (inverse) relationship between the force and the frequency (the velocity of foot motion) is an example of the parametric relationship.

Here are some other examples from different activities:

Activity	Parameter	Force	Velocity	Relationship
Rowing, kayaking, canoeing	Blade area of a row or a paddle	Applied to the row or to the paddle	The blade with respect to the water	Inverse (negative)
Uphill/ downhill ambulation	Incline/decline	At takeoff	Ambulation	Same
Throwing	Weight of the implement	Exerted upon the implement	Implement at the release	Same
Standing vertical jump	Modified body weight. Weight added (waist belt) or deducted (a load is fixed to a rope that is wound around a pulley and fixed to a harness worn by the athlete)	At takeoff	Body at the end of takeoff	Same

Note that all relationships are negative (inverse)—the higher the force, the lower the velocity.

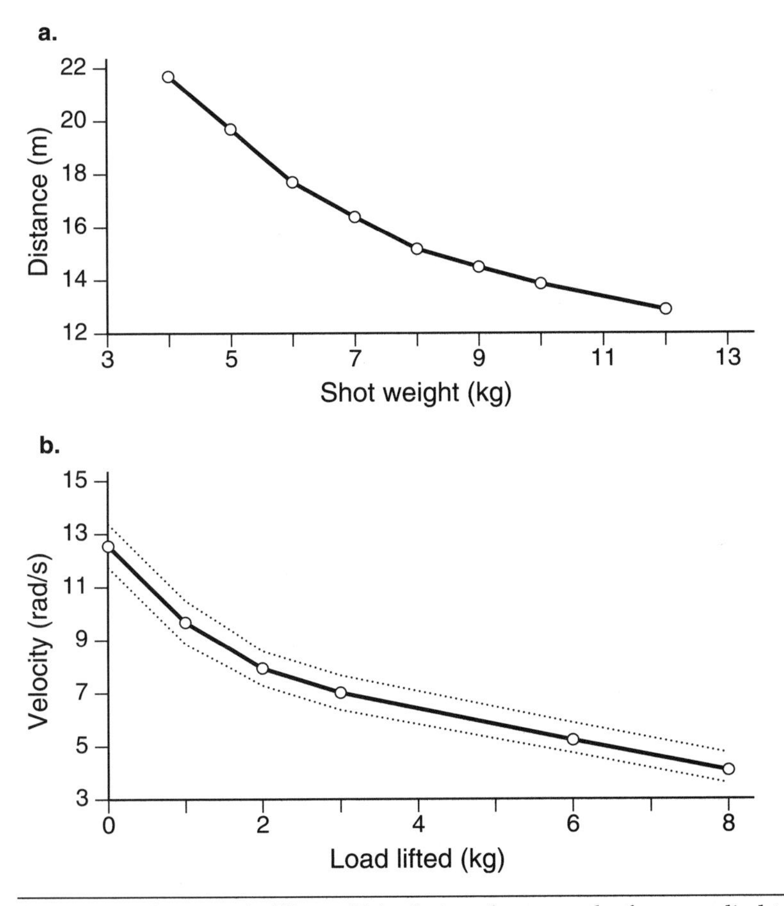

Figure 2.1 Parametric (V_m vs. F_m) relations between the force applied to an implement and implement velocity. (a) The relationship between shot weight and throwing distance. The shot mass varied from 4.0 to 12 kg. Shots were put from a standing position by one subject. In this experiment, the shot weight is the magnitude of the force applied to the shot at the instant of release. At this instant the shot velocity is maximal and, if the velocity is not changing in direction, the acceleration equals zero. The throwing distance, provided that the point of release and the release angle do not vary, is the function of the release velocity. Thus, the relationship between the shot weight and the throwing distance represents (approximately) the parametric force-velocity relationship. Note. From *The Use of Shots of Various Weight in the Training of Elite Shot Putters* by V.M. Zatsiorsky and N.A. Karasiov, 1978. Moscow: Central Institute of Physical Culture. (b) The relationship between the load lifted during shoulder flexion with the arm extended and the maximal lifting velocity V_m. One hundred subjects; average values and standard deviations. Note. From "Relationships Between the Motor Abilities, Part 1" by V.M. Zatsiorsky, Yu. I. Smirnov, and N.G. Kulik, 1969, *Theory and Practice of Physical Culture*, 31(12), pp. 35-48. Reprinted by permission from the journal.

Nonparametric Relationships

Each point on a parametric curve (V_m-F_m) corresponds to the maximal performance at the given value of the motor task parameter (such as object weight, external resistance, distance). Among these performances, there are peak values such as the highest F_m or V_m. These achievements, the highest among the maximal, are termed *maximum maximorum performance* (velocity). The symbols P_{mm}, V_{mm}, and F_{mm} are used to represent them. These levels can be achieved only under the most favorable conditions. For instance, V_{mm} can be attained only if the external mechanical resistance is minimal and the movement time is short (e.g., in the throwing of light objects or in the sprint dash), and F_{mm} can be attained only if the external resistance is sufficiently high.

The relationship between P_{mm} (V_{mm}, F_{mm}), on the one hand, and P_m (V_m, F_m, T_m), on the other, is called the *maximal nonparametric relationship*, or simply the *nonparametric relationship*. The following performance pairs are examples of nonparametric relations:

- The maximal result in a bench press (F_{mm}) and the throwing distance of putting 7- or 4-kg shots (P_m or V_m)
- The maximum maximorum force in a leg extension and the height of a standing jump.

Nonparametric relationships, unlike parametric ones, are typically positive. For instance, the greater the value of F_{mm}, the greater the value of the V_m. The magnitude of this correlation depends on the parameter value of the specific motor task (Figure 2.2). The correlation between maximum maximorum values F_{mm} and V_{mm} is close to zero.

When considering the training of maximal muscular strength, you should distinguish between F_{mm} and F_m.

■ Nonparametric Relationships

A swim coach wants to determine the importance of dryland strength training for her athletes. In order to solve this problem, she measures (a) the maximal force (F_{mm}) produced by the athletes in a specific stroke movement against high resistance and (b) swimming velocity.

She assumes that if the correlation between the two variables is high, then the F_{mm} values are important and it is worthwhile spending the effort and time to enhance maximal force production. If the correlation is low (i.e., the strongest athletes are not the fastest ones), there is no reason to train for maximal strength. Other abilities such as muscular endurance and flexibility are more important.

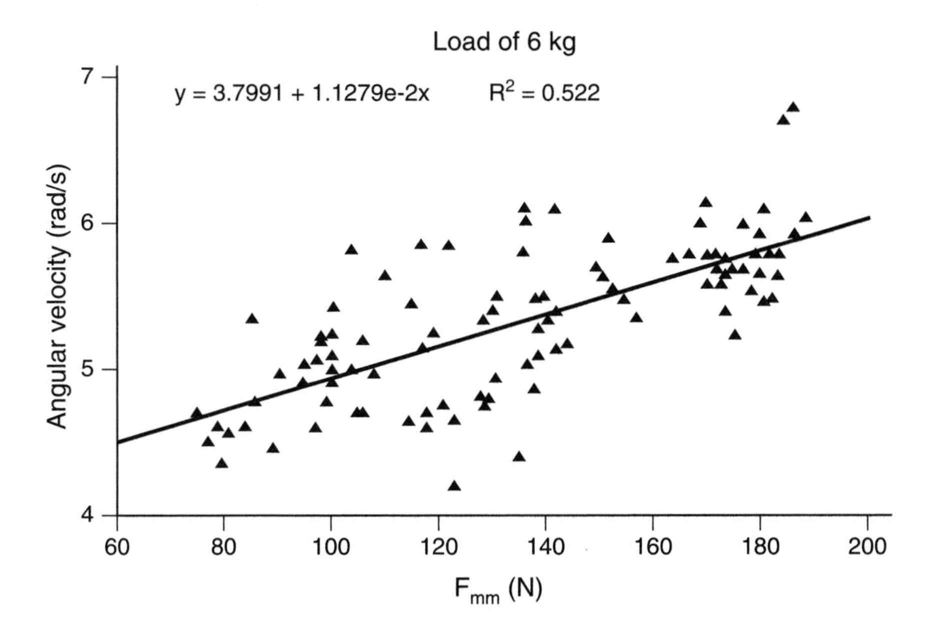

Figure 2.2 Nonparametric relationships between the maximum maximorum force (F_{mm}) and the velocity of shoulder flexion (V_m) with arm extended. Load (a dumbbell) of 6 kg in the hand; 100 subjects. Compare with Fig. 2.12. Note. The data are from *Motor Abilities of Athletes* (p. 46) by V.M. Zatsiorsky, 1969, unpublished doctoral dissertation, Central Institute of Physical Culture, Moscow.

The coach finds that the correlation between F_{mm} and swimming velocity is significant. The better swimmers generate larger forces in specific movements. This is an example of a nonparametric relationship.

Defining Muscular Strength

Strength, or muscular strength, is the ability to generate maximum maximorum external force, F_{mm}. Recall that in mechanics and physics, force is defined as an instantaneous measure of the interaction between two bodies. Force manifests itself in two ways: Either the movement of a body is changed, or the body is deformed, or both. Force is a vector quantity. It is characterized by (a) magnitude, (b) direction, and (c) point of application. Since force is an instantaneous measure and all human movements are performed over a certain span of time, the entire force-time continuum, not just the force at a given instant of time, is typically what interests coaches and athletes.

Many different forces exist in athletic movements. In biomechanics, they are divided into two groups: *internal* forces and *external* forces. A force exerted by one constituent part of the human body on another part is termed an internal force. Internal forces include bone-on-bone forces and tendon-to-bone forces, among others. The forces acting between an athlete's body and the environment are called external forces. Thus, according to this definition of strength, only external forces are regarded as a measure of an athlete's strength.

It is well known that an active muscle exerts force on the bone while

- shortening (*concentric* or *miometric* action),
- lengthening (*eccentric* or *pliometric*), or
- remaining the same length (*static* or *isometric*).

Note that *metric* means length, *mio* means less, *pleio* (*plio-*) means more, and *iso* means same or constant. In the U.S. *plyometrics* has become a common spelling, with *pliometrics* an alternative. Disregarding the differences between muscle force (force developed by a muscle) and muscular strength (maximal force exerted on an external body), this simple classification can be used to discern variations of muscular strength.

In another sense, strength can be defined as the ability to overcome or counteract external resistance by muscular effort. In the case of concentric muscle action, resistance forces act in the direction opposite to the motion, whereas in eccentric action, the external forces act in the same direction as the motion.

■ *What Is Muscular Strength?*

A subject was asked to flex an elbow joint with maximal effort to generate the highest possible force and velocity against different objects. The objects included a dime, a baseball, a 7-kg shot, and dumbbells of different weights, including one that was too heavy to lift. The maximal forces (F_m) applied to the objects were measured and found to be unequal.

The question: Which of the F_m values represents muscular strength?

The answer: According to the definition given, the highest one. The F_{mm}, not F_m, is the measure of muscular strength.

Determining Factors: Comparison Across Tasks

If, in different attempts, all body parts move along the same trajectory or very similar trajectories, we say that the *motion* itself is the same regardless of differences in such elements as time and velocity. So, by definition, a motion is determined only by the geometry of movement, not by its kinematics or kinetics. For instance, a snatch (one of the lifts in Olympic style weight lifting, in which the barbell is lifted from the floor to over the head in one continuous motion) with a barbell of different weights is one motion, while the takeoff in a vertical jump with or without an additional load is a second motion.

Maximal forces exerted by an athlete in the same motion, for instance in the leg extension of the previous examples, are dissimilar if conditions are changed. The two types of factors that determine these differences are extrinsic (external) and intrinsic (internal).

Extrinsic Factors and the Role of Resistance

Force is the measure of the action of one body against another, and its magnitude depends on the features and movements of both bodies in action. The force exerted by an athlete on an external body (e.g., a free weight, a throwing implement, the handle of an exercise machine, water in swimming and rowing) depends not only on the athlete but also on external factors.

To judge the role of external resistance, imagine an athlete who exerts maximal force (F_m) in a leg extension such as squatting. Two experimental paradigms are employed to measure the external resistance. In the first case, the maximal isometric force (F_m) corresponding to different degrees of leg extension is measured. Many researchers have found that the correlation between the force F_m and leg length (i.e., the distance from the pelvis to the foot) is positive: If the leg extends, the force increases (Figure 2.3, curve A; see also Figure 1.3). Maximum maximorum force (F_{mm}) is achieved when the position of the leg is close to full extension. This is in agreement with everyday observations—the heaviest weight can be lifted in semisquatting, not deep squatting, movements.

However, if the leg extension force is registered in a dynamic movement such as a takeoff in jumping, the dependence is exactly the opposite (Figure 2.3, curve B). In this case, maximal force is generated in the deepest squatting position. The correlation of F_m to leg length, then, is negative. Here the mechanical behavior of a support leg resembles the behavior of a spring; the greater the deformation (i.e., knee bending), the greater the force. Remember that in both experimental conditions (isometric and jumping takeoff), the athlete is making maximal effort. Thus, both the magnitude of F_m and the correlation of F_m to leg length (positive

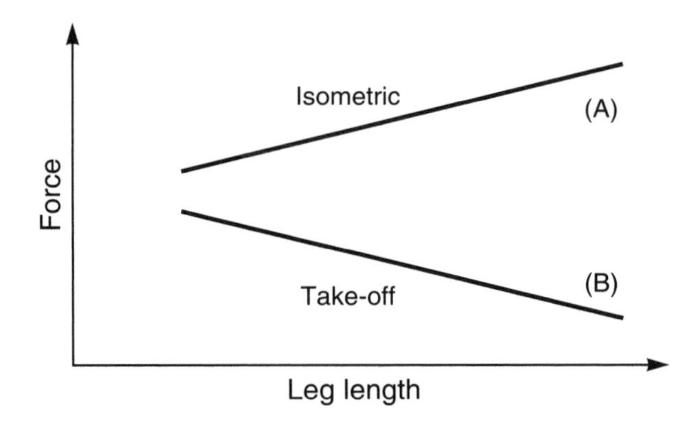

Figure 2.3 Relationship between maximal force in leg extension and body position (leg "length"). (A) Isometric testing. (B) Force generated as the leg extends during a takeoff. See also Figures 1.3 and 2.23 (Leg extension force).

or negative) are changed because the type of resistance changes. In the first case the resistance is the immovable obstacle and in the second it is the weight and inertia of the athlete's body.

Mechanical Feedback

All strength exercises, depending on the type of resistance, can be separated into those with and those without *mechanical feedback*. Let's consider, for instance, a paddling movement in water. In hydrodynamics, the force applied to water is proportional to the velocity squared ($F = kV^2$). However, the oar's velocity is the result of an athlete's efforts, an external muscular force. The chain of events is represented in Figure 2.4. Here, active muscular force leads to higher oar velocity, which in turn increases water resistance. Then, to overcome the increased water resistance, the muscle force is elevated. Thus, increased water resistance can be regarded as an effect of the high muscular force (*mechanical feedback*).

Imagine a different example, that of an individual pushing an already moving heavy truck. Regardless of all the force applied by the person, the truck moves with the same velocity. The human's muscular efforts result in no change in the truck's movement (no mechanical feedback).

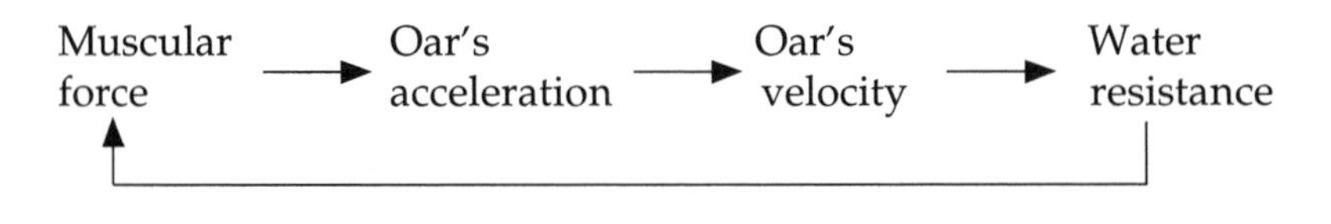

Figure 2.4 Mechanical feedback loop.

Sport movements usually involve mechanical feedback: The movement, as well as resistance, is changed as a result of an athlete's force application. Mechanical feedback is absent only in the performance of isometric exercises and in work with isokinetic devices.

With isokinetic devices, the velocity of limb movements around a joint is kept constant. The resistance of the device is equal to the muscular force applied throughout the range of movement. The maximal force F_m is measured in dynamic conditions, provided that the preset velocity has been attained by the moving limb.

Types of Resistance

Because of the specific requirements of strength exercises, selecting the proper class of mechanical resistance equipment is important in training. The equipment typically used in resistance training programs can be categorized according to the type of resistance involved.

In resistance based on *elasticity*, the magnitude of force is determined by the range of displacement. The length of an object with ideal elasticity increases in proportion to the force applied. The formula is $F = k_1D$, where F is force, k_1 is a coefficient (stiffness), and D is displacement (deformation). In other words, the greater the range of motion (e.g., the deformation of a spring, stretch cord, or rubber band), the higher the exerted muscular force.

Another type of resistance is based on *inertia*. A movement follows Newton's second law of motion: $F = ma$, where *m* is mass and *a* is acceleration. The force is proportional to the mass (inertia) of the accelerated body and its acceleration. As the body mass is typically selected as a parameter of a motor task, the force determines the acceleration. Because of gravity and friction, however, it is difficult to observe movement in which the resistance is formed only by inertia. The motion of a billiard ball is one example.

In science, movement against inertial resistance is studied by using an *inertia wheel*, or pulley, that rotates freely around an axis perpendicular to its surface plane. A rope is wound repeatedly around the pulley and a subject then pulls the rope; this force exerted by the subject in turn rotates the pulley and does mechanical work. With this device, the potential energy of the system is constant and all mechanical work, except small frictional losses, is converted into kinetic energy. By varying the mass (or moment of inertia) of the wheel, we can study the dependence of exerted muscular force, particularly F_m, upon the mass of the object. The results are shown in Figure 2.5.

If the mass of an accelerated object is relatively small, the maximal force exerted by an athlete depends on the size of the mass (see zone A in Figure 2.5). It is impossible to exert a large F_m against a body of small mass. For instance, it is unrealistic to apply a great force to a coin. If the mass of an object is great, however, the F_m depends not on the body mass but only on the athlete's strength (Figure 2.5, zone B).

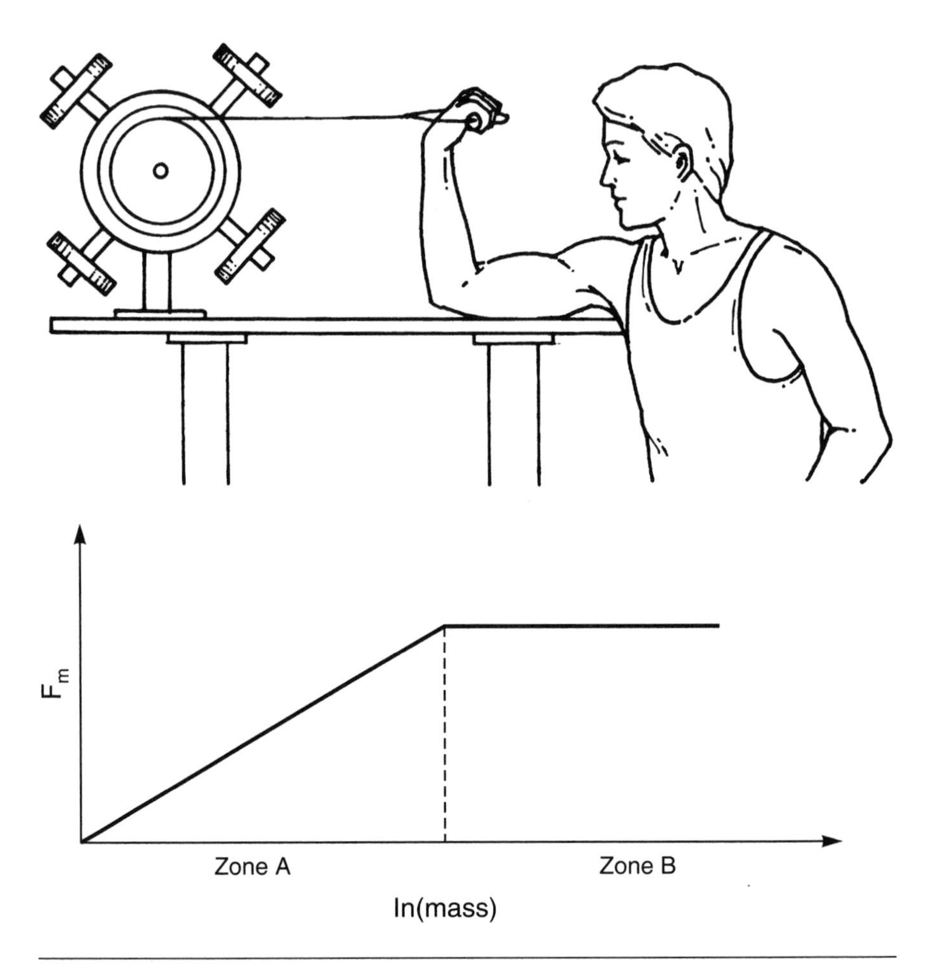

Figure 2.5 The inertia wheel device (top) and the dependence of maximal exerted force F_m on the mass of the moving object (bottom). Scale on the abscissa axis is logarithmic. Note. From *Motor Abilities of Athletes* by V.M. Zatsiorsky, 1966, Moscow: Fizkultura i Sport. Reprinted by permission from the publisher.

An example from sport training shows the relationship between mass and force. When objects of different masses are thrown (e.g., shots 1.0–20.0 kg are used in training), the force applied to the light shots is relatively small and heavily influenced by the shot mass (zone A). The force exerted on the heavy shots, however, is determined only by the athlete's strength (zone B).

Resistance can also be based on *weight*. The formula is $F = W + ma$, where W is the weight of the object and a is the vertical acceleration. If a is zero (the object is at rest or in uniform motion), the force equals the object weight. When exercising with free weights, an athlete needs to fix the barbell in a static position. Typically, it is not feasible to relax before and immediately after the effort as is possible for a motion against other types of resistance (for instance in a swimmer's stroke). All exercises in which athletes move their own bodies (gymnastic strength exercises) are classified as having this type of resistance.

If a body is accelerated by muscular force, the direction of the acceleration does not coincide with the direction of this force except when the movement is vertical. Rather, it coincides with the direction of the resultant force, which is a vector sum of the muscular force and the force of gravity. Since gravity is always acting downward, the athlete should compensate for this action by directing the effort higher than the desired movement direction. For instance, in shot putting the direction of the shot acceleration does not coincide with the direction of the athlete's force applied to the shot (Figure 2.6). The same is true for jump takeoffs.

■ *Why Is Strength Training Vital for Sprinters and Jumpers?*

Because body weight (during the upward takeoff motion) and body mass (during both the horizontal and vertical push-offs) provide high resistance. If you practice a leg extension without any external resistance, the strength training will be of small value, since there is no positive relationship between the maximum maximorum force (F_{mm}) and the maximum maximorum velocity (V_{mm}).

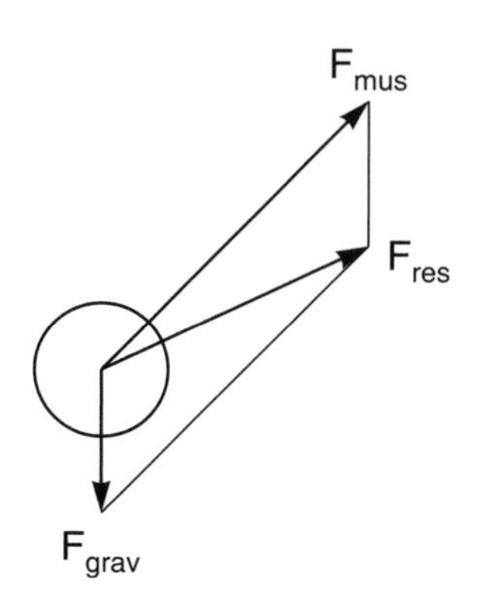

Figure 2.6 Muscular (F_{mus}) and gravity (F_{grav}) forces applied to a shot. Shot acceleration coincides in direction with the resultant forces (F_{res}) but not with F_{mus}.

Hydrodynamic resistance predominates in water sports such as swimming, rowing, and kayaking. Force in this case depends on the velocity squared: $F = k_2V^2$, where V is the velocity relative to water and k_2 is a coefficient of hydrodynamic resistance.

It is difficult to model this type of resistance on land. Thus the selection of proper strength or dryland training in water sports is a special problem. The use of weights or elastic resistance is not a satisfactory solution. While performing a stroke in the water, the athlete relaxes immediately before and after the stroke and also exerts maximal force against the water resistance at a time when the maximal velocity is achieved. These two features are both unattainable with springs and free weights.

With some training devices the resistance is provided by *viscosity*. Here the exerted muscular force is proportional to the movement velocity, $F = k_3V$. These exercise machines are mainly used as a substitute for natural water conditions and for dryland training in water sports.

■ *Selection of Dryland Exercises for Swimmers*

A swim coach explored several types of training devices for dryland training. Lying in a prone position on a couch, the athletes initiated a stroke pattern against provided resistance. First they used extensible rubberlike bands. However, during this exercise the pulling force inevitably increased from the beginning to the end of the pull. This movement pattern is not similar to the customary stroke. Then the swimmers used a weightlifting exercise machine with a pulley to pull a rope attached to a load. The resistance was almost constant over the range of the pull, but they couldn't relax their muscles at the end of the motion. Their arms were forcibly jerked in the reverse direction. Finally the athletes used training devices with friction resistance (or hydrodynamic resistance). These provided either constant resistance (friction devices) or resistance proportional to the pull velocity (hydrodynamic exercise machines), which mimicked water resistance. The resemblance, however, was far from ideal; during the natural stroke, the resistive force is proportional to the squared values of the hand velocity with respect to the water.

Intrinsic Factors

The strength that an athlete can exert in the same motion depends on several variables: velocity, body position, and direction of movement. The cause of muscular strength is, obviously, the activity of individual muscles. The variables just mentioned also determine the force output of single muscles. However, the relationship between the activity of specific muscles and muscular strength (e.g., in lifting a barbell) is not straightforward. Muscular strength is determined by the concerted activity of many muscles. Active muscles produce a pulling effect on the bones in a straight line. But the translatory action of muscle forces also induces a rotatory movement in the joints. As various muscles are inserted at different distances from the joint axes of rotations, their rotatory actions (moments of force) are not in direct proportion to the force developed by muscles. The rotatory movements in several joints are coordinated so as to produce the maximal external force in a desired direction, such as the vertical direction required to lift a barbell. Thus, complicated relationships exist between muscle force (force exerted by a given muscle) and muscular strength (maximal external force). Regardless of these differences, many facets of muscular biomechanics and the physiology of isolated muscles are manifested in the complex movements involving numerous muscles.

Time

It takes time to develop maximal force for a given motion (Figure 2.7).

The time to peak force (T_m) varies with each person and with different motions; on average, if measured isometrically, it is approximately 0.3 to 0.4 s. (Typically, the time to peak force is even longer than 0.4 s. However, the final increase in force is very small (< 2–3% of F_m) and force output

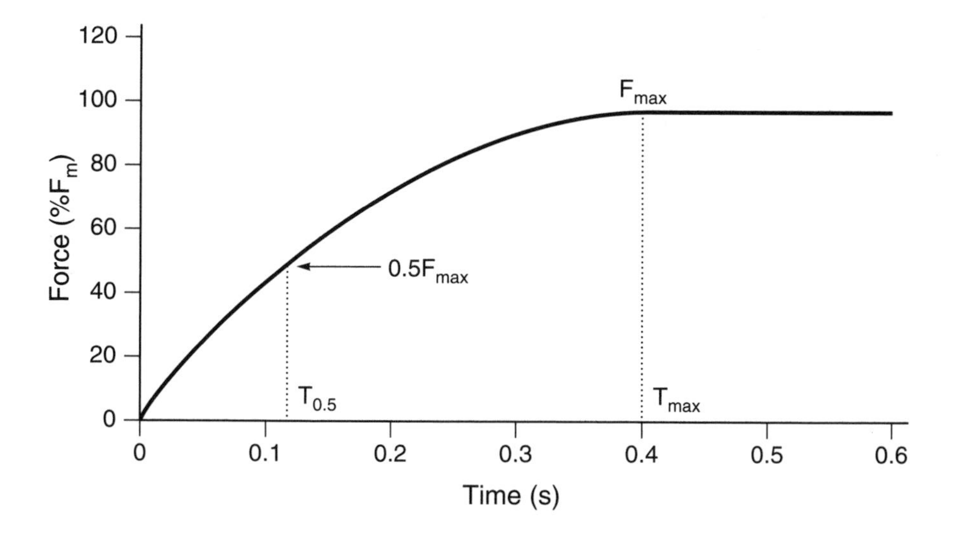

Figure 2.7 Development of maximal muscular force over time. T_m is the time to F_m , $T_{0.5}$ is the time to 1/2 of F_m.

begins to fluctuate, preventing a precise determination of the time to peak force. In practice, the final portion of the force-time curve is usually disregarded.) The time for maximal force development can be compared with the time typically required by elite athletes to perform several motions:

Motion	*Time (s)*
Takeoff	
Sprint running	0.08–0.10
Long jump	0.11–0.12
High jump	0.18
Delivery	
Javelin	0.16–0.18
Shot put	0.15–0.18
Hand takeoff	
Horse vaulting	0.18–0.21

It is easy to see that the time of motion is less than T_m in all examples given. Because of their short durations, the maximal possible force F_{mm} cannot be attained during the performance of these motions.

As the resistance decreases and the motion time becomes shorter, the difference between F_m (the maximal force reached in a given condition) and F_{mm} (the highest among the maximal forces attained in the whole range of the tested conditions) increases (Figure 2.8).

The difference between F_{mm} and F_m is termed the *explosive strength deficit* (ESD). By definition:

$$\text{ESD}(\%) = 100 \cdot (F_{mm} - F_m)/F_{mm}$$

ESD shows the percentage of an athlete's strength potential that was not used in a given attempt. In movements such as takeoffs and delivery phases in throwing, ESD is about 50%. For instance, among the best shot-putters during throws of 21.0 m, the peak force F_m applied to the shot is in the range of 50 to 60 kg. The best results for these athletes in an arm extension exercise (F_{mm}, bench press) are typically about 220 to 240 kg, or 110 to 120 kg for each arm. Thus, in throwing, they can only use about 50% of F_{mm}.

In principle, there are two ways to increase the force output in explosive motions—to increase F_{mm} or decrease ESD. The first method brings good results at the beginning of sport preparation. If a young shot-putter improves achievement in, say, bench press from 50 to 150 kg and also pays proper attention to the development of other muscle groups, this athlete has a very strong basis for better sport performance in shot

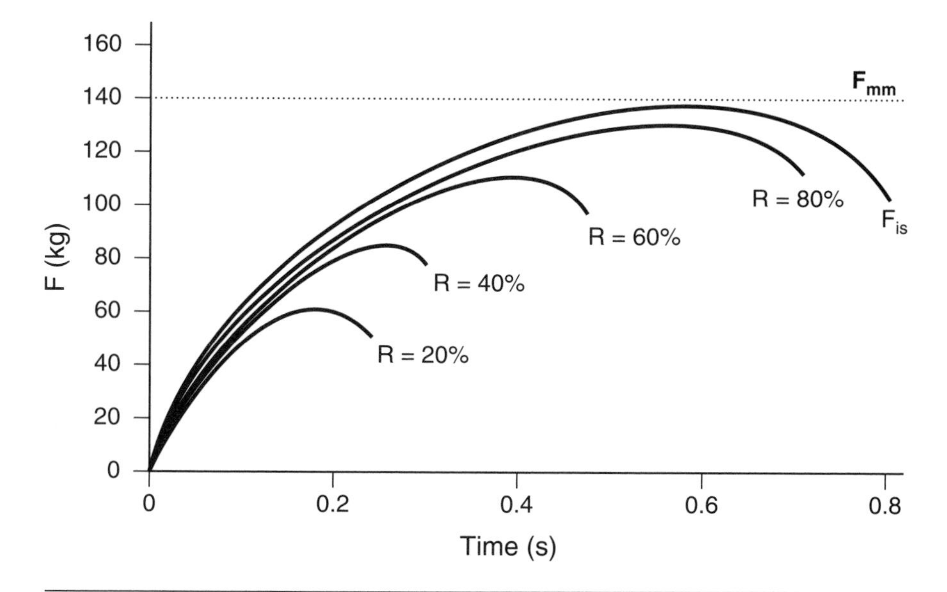

Figure 2.8 Force-time histories of a leg extension against different levels of resistance. The subject was asked to perform the effort in an explosive way, i.e., as quickly and strongly as possible. The magnitude of weights, i.e., resistance (R), varied from 20% to 80% of F_{mm}. F_{mm} was determined in the isometric condition without any restrictions with respect to time. Force-time curve for an explosive isometric effort is also shown. Note. Adapted from *Special Strength Training in Sport* (p. 76) by Yu. V. Verchoshansky, 1977, Moscow: Fizkultura i Sport. Adapted by permission from the author.

putting. This is not necessarily valid, however, for a bench press gain from 200 to 300 kg. In spite of efforts devoted to making such a tremendous increase, the shot-putting result may not improve. The reason for this is the very short duration of the delivery phase. The athlete simply has no time to develop maximal force (F_{mm}). In such a situation, the second factor, explosive strength, not the athlete's maximal strength (F_{mm}), is the critical factor. By definition, *explosive strength* is the ability to exert maximal forces in minimal time.

Let's compare two athletes, A and B, with different force-time histories (Figure 2.9). If the time of motion is short (i.e., in the *time deficit zone*), then A is stronger than B. The situation is exactly opposite if the time of the movement is long enough to develop maximal muscular force. Training of maximal strength cannot help athlete B improve performance if the motion is in the time deficit zone.

When sport performance improves, the time of motion turns out to be shorter. The better an athlete's qualifications, the greater the role of the rate of force development in the achievement of high-level performance.

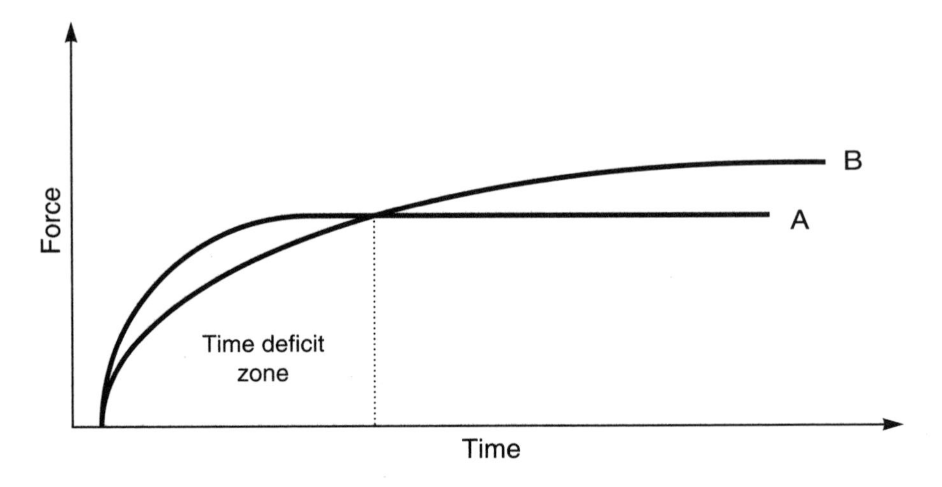

Figure 2.9 Force-time histories of two athletes, A and B. In the time deficit zone, A is stronger than B.

Several indices are used to estimate explosive strength and the rate of force development (see Figure 2.7 for the key to the symbols). These are

(a) *Index of explosive strength* (IES)

$$IES = F_m/T_m,$$

where F_m is the peak force and T_m is the time to peak force.

(b) *Reactivity coefficient* (RC)

$$RC = \frac{F_m}{T_m \cdot W}$$

where W is an athlete's weight (or weight of an object). RC is typically highly correlated with jumping performances, especially with body velocity after a takeoff.

(c) *Force gradient,* also called the *S-gradient* (*S* for start)

$$\text{S-gradient} = F_{0.5} / T_{0.5}$$

where $F_{0.5}$ is one half of the maximal force F_m and $T_{0.5}$ is the time to attain it. S-gradient characterizes the rate of force development at the beginning phase of a muscular effort.

(d) *A-gradient* (*A* for acceleration)

$$\text{A-gradient} = F_{0.5} / (T_{max} - T_{0.5})$$

A-gradient is used to quantify the rate of force development in the late stages of explosive muscular efforts.

F_m and the rate of force development, particularly the S-gradient, are not correlated. Strong people do not necessarily possess a high rate of force development.

■ *Defining a Training Target: Strength or Rate of Force Development?*

A young athlete began to exercise with free weights, performing squats with a heavy barbell. At first he was able to squat a barbell equal to his body weight (BW). His performance in a standing vertical jump was 50 cm. After 2 years, his achievement in the barbell squats was 2 BW, and the vertical jump increased to 80 cm. He continued to train in the same manner and after 2 more years was able to squat with a 3-BW barbell. However, his jump performance was not improved because the short takeoff time (the rate of force development) rather than maximal absolute force became the limiting factor.

Many coaches and athletes make a similar mistake. They continue to train maximal muscle strength when the real need is to develop rate of force.

Velocity

The force-velocity relationship is a typical example of the parametric relations described in general earlier in the discussion of maximal muscular performance. Motion velocity decreases as external resistance (load) increases. For instance, if an athlete throws shots of different weights, the throwing distance (and initial velocity of the implement) increases as shot weight decreases. Maximum force (F_{mm}) is attained when velocity is small; and, inversely, maximum velocity (V_{mm}) is attained when external resistance is close to zero (Figure 2.10).

Experiments carried out on single muscles in laboratory conditions yield the well known force-velocity curve (Figure 2.11), which can be described by the hyperbolic equation

$$(F + a)(V + b) = (F_{mm} + a)b = C$$

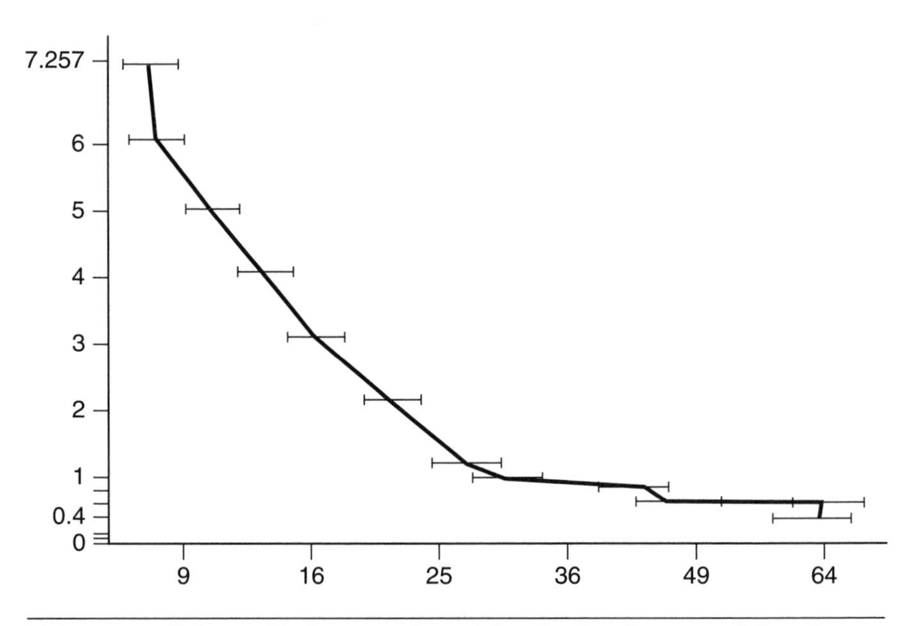

Figure 2.10 Relationship between the weight of an implement and throwing distance (velocity at release). Athletes (n = 24) performed overhand throwing of shots of different weights from a standing position. Ordinate, weight of shots (kg); abscissa, square roots of the throwing distances (meters). Note. From "Force-Velocity Relationships in Throwing (as Related to the Selection of the Training Exercises)" by V.M. Zatsiorsky and E.N. Matveev, 1964, *Theory and Practice of Physical Culture*, **27**(8), pp. 24-28. Reprinted by permission from the journal.

where

F is the force;

V, velocity of muscle shortening;

F_{mm}, maximal isometric tension of that muscle;

a, a constant with dimensions of force;

b, a constant with dimensions of velocity; and

C, a constant with dimensions of power.

The force-velocity curve can be considered part of a hyperbolic curve with the axis (external) shown in Figure 2.11. The curvature of the force-velocity graph is determined by the ratio $a{:}F_{mm}$. The lower the ratio, the greater the curvature and the closer the force-velocity relation to an asymptotic hyperbola. Line curvature decreases if $a{:}F_{mm}$ increases. Various main sport movements encompass different parts of the force-velocity curves.

Force-velocity (as well as torque-angular velocity) relationships in human movements are not identical to analogous curves of single muscles because they are a result of the superposition of the force outcome of

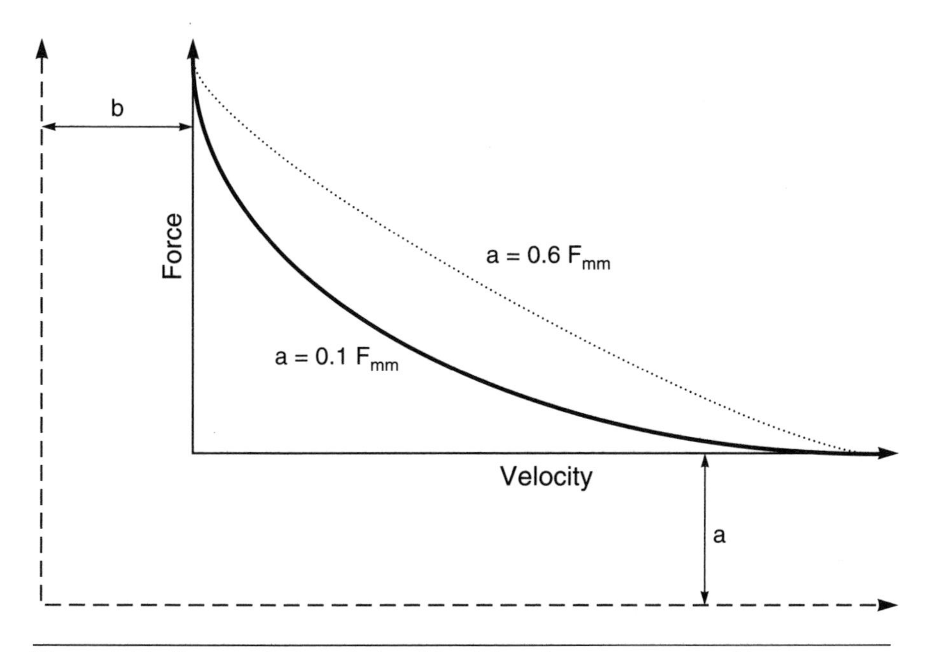

Figure 2.11 Force-velocity relationships. Note the constants a and b.

several muscles possessing different features. Nevertheless, force-velocity curves registered in natural human movements can be considered hyperbolic. This approximation is not absolutely accurate, but the accuracy is acceptable for the practical problems of sport training. The ratio $a{:}F_{mm}$ varies from 0.10 to 0.60. Athletes in power sports usually have a ratio higher than 0.30, while endurance athletes and beginners have a ratio that is lower.

The second factor that may change the hyperbola-like appearance of the force-velocity curve in natural athletic motions is the time required for force development. In fast movements, the time may be too short to develop maximal force, thus distorting the "real" force-velocity curve. To exclude the influence of the time available for force development, experimenters use the *quick release* technique. In this method a subject develops force under isometric conditions with a body segment mechanically locked into position. The lock is then trigger released, barely permitting the subject to perform a movement against the given resistance. In this case, the initial conditions for muscle shortening are determined by the magnitude of force, not the rate or time of force development.

Force-velocity relationships can also be studied with isokinetic devices that keep velocity constant during a movement. However, the velocity range of modern isokinetic equipment is relatively small, preventing the study of very fast movements.

Several consequences of the force-velocity equation are important for sport practice:

1. It is impossible to exert a high force in very fast movements. If an athlete performs the first phase of a movement too fast, the ability to apply great force in the second phase may be somewhat diminished. For instance, too fast a start in lifting a barbell from the floor may prevent an athlete from exerting maximal force in the most advantageous position—when the barbell is near the knees.

2. The magnitudes of force and velocity developed in the intermediate range of the force-velocity curve depend on the maximal isometric force F_{mm}. In other words, an athlete's maximal strength F_{mm} determines the force values that can be exerted in dynamic conditions. The dependence of force and velocity developed in dynamic conditions on the maximal force F_{mm} is greater in movements with relatively high resistance and slow speed (Figure 2.12, a and b). At the same time, there is no correlation between maximal force (F_{mm}) and maximal velocity (V_{mm}). The ability to produce maximal force (i.e., muscular strength) and the ability to achieve great velocity in the same motion are different motor abilities. This is true for extreme areas of the force-velocity curve, while intermediate values depend on the F_{mm}.

■ *Why Do Shot-Putters and Javelin Throwers Pay Different Attention to Heavy Resistance Training?*

In sports such as shot putting and javelin throwing, as well as in throwing in baseball or softball, the motor task is similar—to impart maximal velocity to an implement. Why then do athletes in these sports train differently (and why are their physiques so dissimilar)? Elite shot-putters spend about 50% of their total training time on heavy resistance training, while world-class javelin throwers spend only 15% to 25% of their total training time in the weight room. The reason? Because the implement weights are so different. The shot weight is 7.257 kg for men and 4 kg for women; the javelins weigh 0.8 and 0.6 kg. For top athletes, the velocity of a shot release is nearly 14 m/s, while javelin release velocity is above 30 m/s. These values correspond to different parts of a (parametric) force-velocity curve. The shot-putters need a high F_{mm} because of a high (nonparametric) correlation between maximal strength and the velocity of movement at delivery phase (and similarly, the shot velocity). This correlation is low in javelin throwing. In turn it would be much smaller for a table tennis stroke, since the paddle is very light. And the correlation is zero when the maximal strength (F_{mm}) is compared to the maximal velocity (V_{mm}) of an unloaded arm.

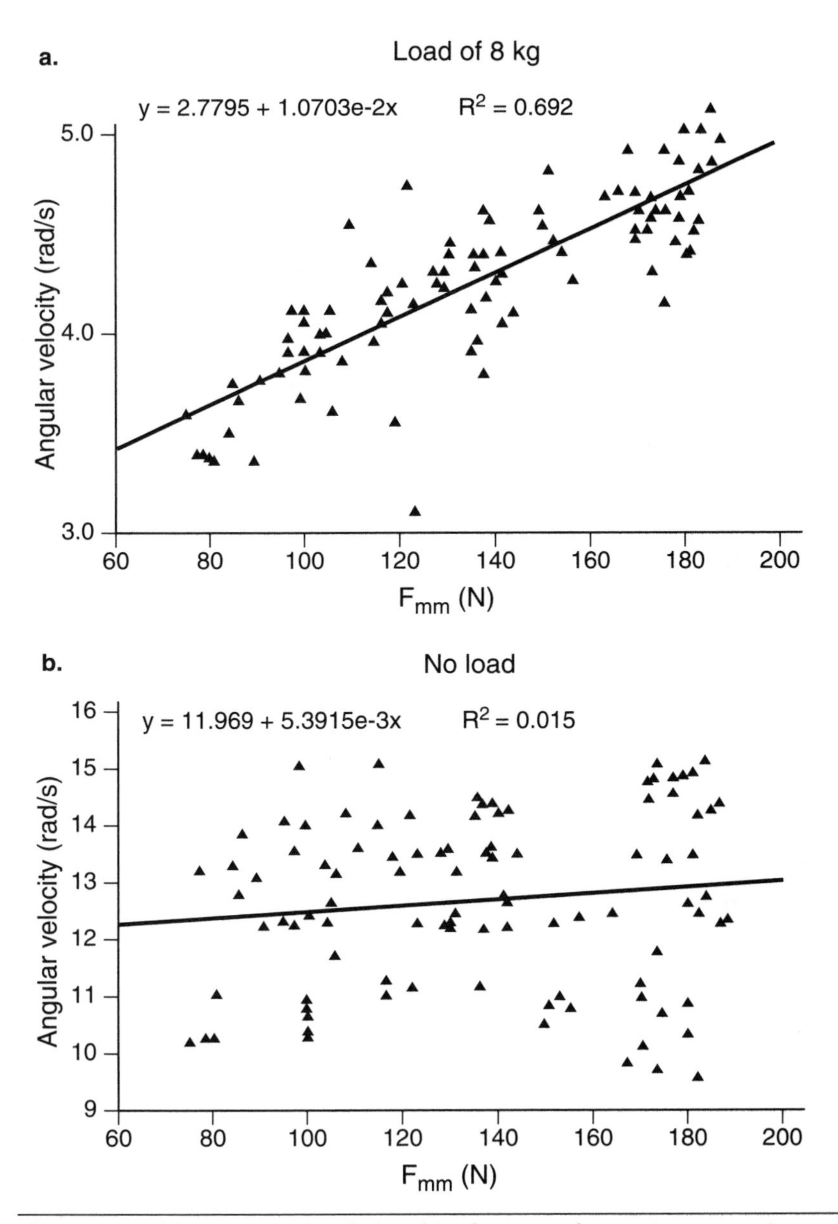

Figure 2.12 Nonparametric relationships between the maximum maximorum force (F_{mm}) and the velocity of shoulder flexion (V_m) with arm extended. Scattergrams of F_{mm} vs. the V_m (a) and V_{mm} (b) are shown. Compare with Figure 2.2. (a) Load (a dumbbell) of 8 kg in the hand; there is a high correlation between F_{mm} and angular velocity (V_m). (b) No load; there is no significant correlation between F_{mm} and V_{mm}. Note. The data are from *Motor Abilities of Athletes* (p. 48) by V.M. Zatsiorsky, 1969, unpublished doctoral dissertation, Central Institute of Physical Culture, Moscow.

3. Maximal mechanical power (P_{mm}) is achieved in the intermediate range of force and velocity. As the velocity of the movement increases, the exerted force decreases and the released energy (work + heat) increases. Efficiency (i.e., ratio of work to energy) achieves its greatest value when the velocity is about 20% of V_{mm} with mechanical power greatest at speeds of about one third of maximum (Figure 2.13).

It may seem surprising that the greatest power value is at a velocity one third the value of maximal velocity (V_{mm}). One should not forget, however, that in the simplest case, power equals force multiplied by velocity:

$$P = \frac{W}{t} = F \cdot \frac{D}{t} = F\left(\frac{D}{t}\right) = F \cdot V$$

where P is power, W is work, F is force, D is distance, t is time, and V is velocity. Since F_m and V_m are inversely related, the power is maximal

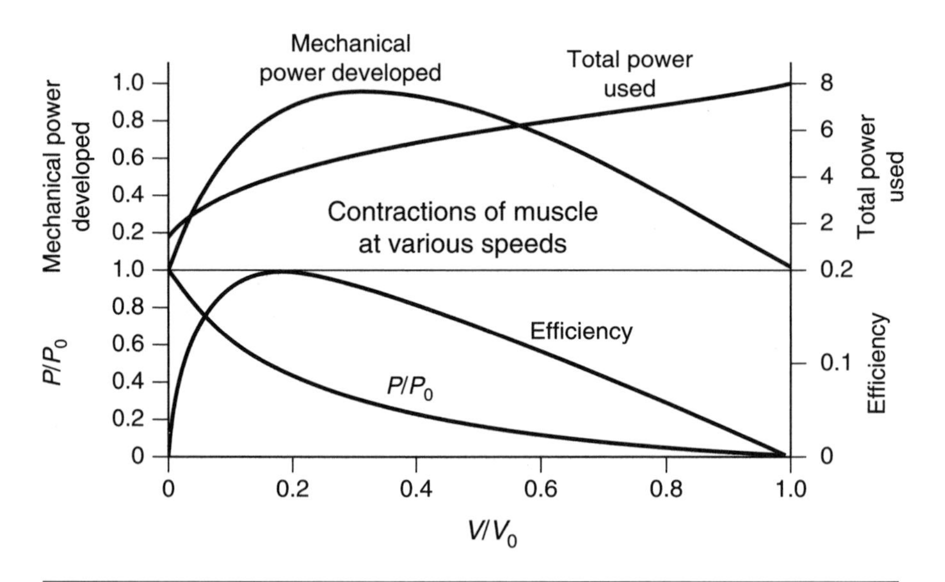

Figure 2.13 Dependence of various movement variables on motion velocity. Abscissa: speed V as a fraction of maximal speed V_0 under zero load (the symbol V_{mm} is used for this quantity throughout this book). Ordinate: (a) Force exerted = P as fraction of maximum force P_0 at zero speed; (b) Efficiency = (mechanical work done)/(total energy used); (c) Mechanical power = PV; (d) Total power used = PV/(efficiency). From experiments performed on isolated muscles and on men. Note. From "The Dimensions of Animals and their Muscular Dynamics" by A.V. Hill, 1950, *Science Progress*, **38**, pp. 209-230. Reprinted by permission from Blackwell Scientific Publications.

when the magnitudes of force and velocity are optimal—about one third of maximal levels of maximal velocity (V_{mm}) and about one half of maximal force (F_{mm}). As a consequence, the maximal power (P_{mm}) equals approximately one sixth of the value that could be achieved if one were able to exert both highest force (F_{mm}) and highest velocity (V_{mm}) simultaneously: $P_{mm} = 1/3\ V_{mm} \cdot 1/2\ F_{mm} = 1/6\ (V_{mm} \cdot F_{mm})$.

This is why the power level is greater when a relatively light shot is put than when a heavy barbell is lifted. For example, the power level is 5,075 W (6.9 horsepower [HP]) in putting a 7.25-kg shot 18.19 m, but only 3,163 W (4.3 HP) during the snatch of a 150-kg barbell. At the same time, the maximal applied force F_m is equal to 513 N for the shot and 2,000 N for the snatch. Though the exerted force is less in shot putting, the exerted power is greater in this case because of the much higher speed of movement.

In some sport movements, it is possible to change the magnitude of external resistance (e.g., cycling gear, rowing paddles). If the final aim in this case is to develop maximal power P_{mm}, it can be achieved with a certain optimal combination of resistance (external force) and cadence (velocity).

Direction of Movement (Pliometrics, Stretch-Shortening Cycle)

Force in the yielding phases of a motion, under conditions of imposed muscle lengthening (*eccentric* or *pliometric* muscle action), can easily exceed the maximal isometric strength of an athlete by up to 50–100%. The same holds true for isolated muscles. The eccentric force for a single muscle may reach a level of up to twice the zero velocity (isometric) force.

Eccentric muscle action. A typical example of eccentric muscle activity can be seen in landing. The force exerted during the yielding phase of landing from a great height can substantially exceed either the takeoff or maximal isometric force. The ground reaction force is typically higher in the first half of the support period (during the yielding phase when the hip, knee, and ankle joints are flexing) than in the second half when the joints extend.

For another example, consider the grip force exerted during the lifting of a heavy barbell. The maximal isometric grip force of male weight lifters, measured with a grip dynamometer, is typically less than 1,000 N and is much lower than the force applied to the barbell. For instance, an athlete lifting a 250-kg barbell applies a maximal instantaneous force of well over 4,000 N to the weight. The force, 2,000 N per arm, is needed to accelerate the barbell. Although the maximal grip strength is only half as high as the force applied to the barbell, the athlete can sustain this great force without extending the grip.

Eccentric forces substantially increase with initial increases in joint movement velocity (and correspondingly the velocity of muscle lengthening) and then remain essentially constant with additional increases in velocity (Figure 2.14). This is mainly true for qualified athletes and in multijoint motions such as the leg extension (according to recently published data, in untrained persons, maximal voluntary torque output during eccentric knee extension or flexion is independent of movement velocity and remains at an isometric level).

If the same external force is exerted concentrically and eccentrically, fewer muscle fibers are activated while the muscle lengthens. Because of this, if the same force is developed, the level of electric activity of muscles (EMG) is lower in exercises with eccentric muscle action. Furthermore, because exercises with eccentric muscle action typically involve high force development, the risk of injury is high—a risk coaches should understand. Even if the eccentric force is not maximal, such exercises (e.g., downhill running) may easily induce delayed muscle soreness, especially in unprepared athletes. The cause of the muscle soreness is damaged muscle fibers. This damage is often considered a normal precursor to the adaptation of muscle to increased use. Conditioning muscle reduces the amount of injury.

Reversible muscle action. Eccentric muscle actions are as natural in human movements as are concentric actions. Many movements consist of eccentric (stretch) and concentric (shortening) phases. This stretch-shortening cycle is a common element of many sport skills and is referred to as the *reversible action* of muscles.

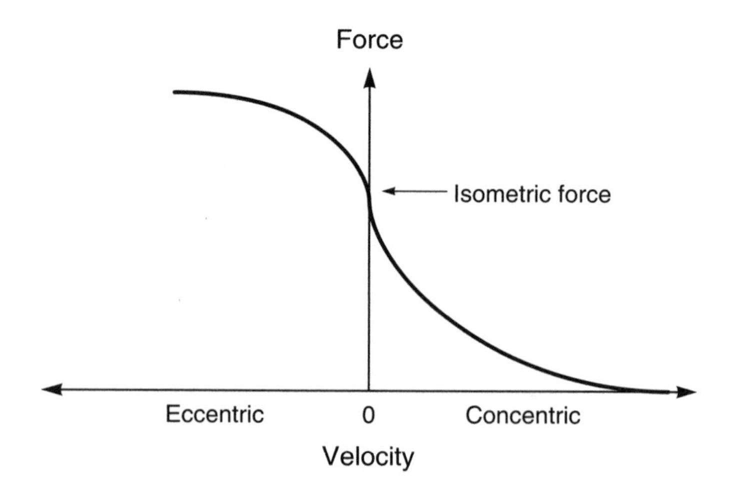

Figure 2.14 Force-velocity curve for concentric and eccentric muscle action.

If a muscle shortens immediately after a stretch

- force and power output increases (Figure 2.15), and
- energy expenditure decreases.

Thus, muscles can produce greater mechanical force and power while utilizing less metabolic energy. Active muscles are typically prestretched to enhance force (power, velocity) output in sport movements. The windup movement in throwing serves as an example.

Reversible muscle action is an innate part of some movements, such as the landing and takeoff in running ("a spring in the leg"); in other movements, such actions must be learned. Since many sport movements are highly complex and executed in a very brief time, even some elite athletes fail to perform this reversed muscle action correctly (Figure 2.16).

Increased force is exerted in the shortening phase of a stretch-shortening cycle for several reasons. At the peak of the cycle, that is, at the moment of transition from lengthening to shortening, the force is developed in isometric conditions; thus the influence of high velocity is avoided, and

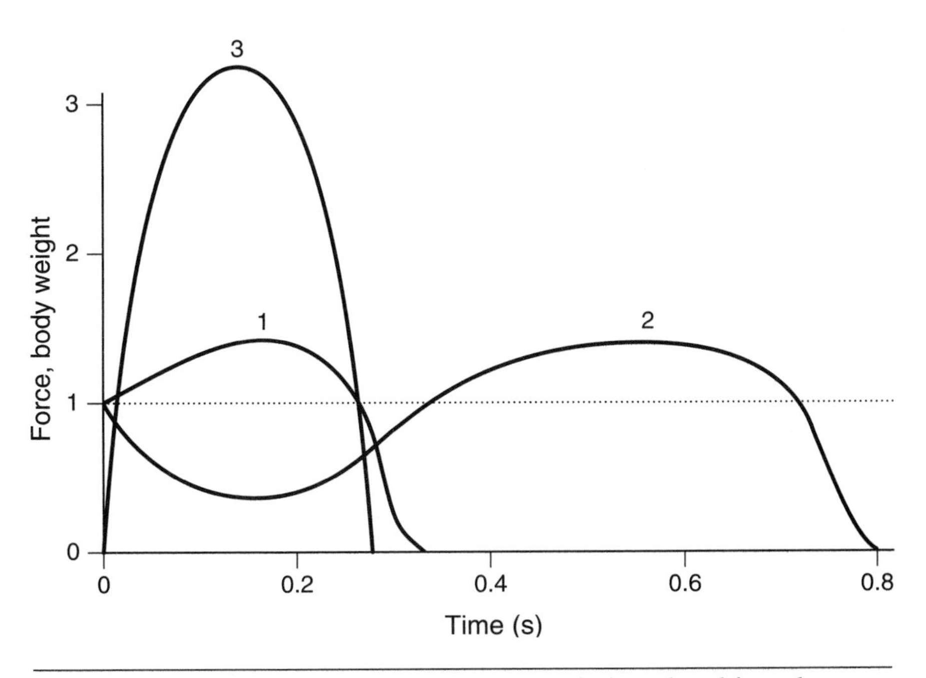

Figure 2.15 The ground reaction force (body weights) produced from three types of vertical jumps: (1) standing jump from a deep squatting position (height of jump was 0.67 m); (2) countermovement jump with deep squatting (0.74 m); (3) drop jump from 40 cm height (0.81 m). The subject was an experienced triple jumper. Note. From *Special Strength Training in Sport* by Yu.V. Verchoshansky, 1977, Moscow: Fizkultura i Sport. Adapted by permission from the author.

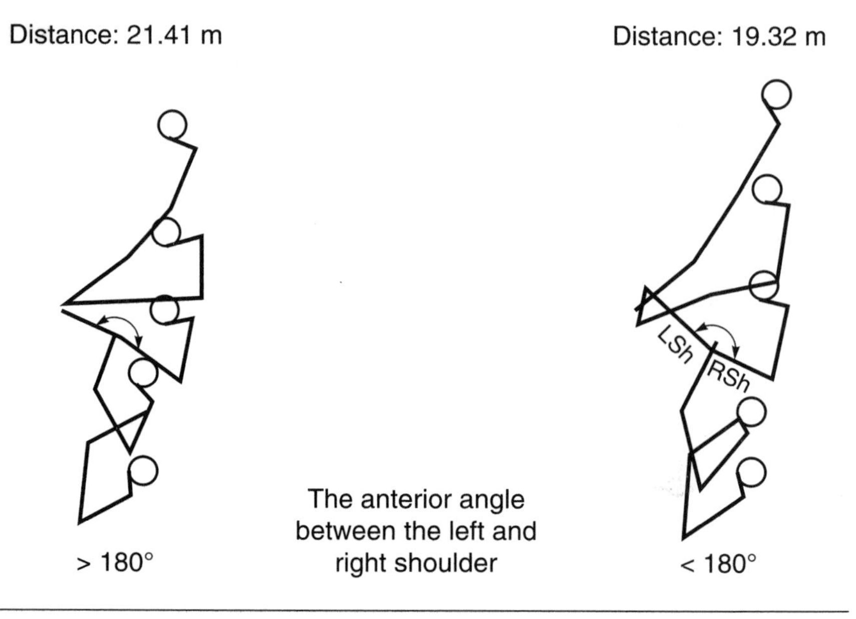

Figure 2.16 Stick figure showing two attempts in shot putting with different results (top view). LSh and RSh are the left and right shoulders. In the successful attempt (result 21.41 m), the athlete managed to stretch the muscles of the shoulder girdle prior to delivery. In the less successful attempt (19.32 m), this element of technique was not properly executed. Note. Adapted from *Sport Technique of Elite Shot Putters* by K. Bartoniets, 1990, June 7-9, paper presented at The First International Conference: Technique in Athletics, Köln.

F_{mm} rather than F_m is exerted. Since the force begins to rise in the eccentric phase, the time available for force development is greater. Countermovement jumps (not drop jumps) are evidence of such an occurrence.

Apart from these mechanisms, two other factors influence the outcome of movements with reversible muscle action: peripheral, or muscle and tendon elasticity, and central (neural), or reflex action.

Muscle and tendon elasticity. Elasticity plays a substantial role in enhancing the motor output in sport movements. If a tendon or active muscle is stretched, the elastic energy is stored within these biological structures. This deformation energy is recoiled and used to enhance motor output in the concentric phase of the stretch-shortening cycle. According to physical principles, the magnitude of the stored energy is proportional to the applied force and the induced deformation. Since muscle and tendon are arranged in series, they are subjected to the same force, and the distribution of the stored energy between them is in this case only a function of their deformation. The deformation, in turn, is a function of muscle or tendon *stiffness* (or its inverse value, *compliance*).

The stiffness of a tendon is constant, while the stiffness of muscles is variable and depends on the forces exerted. The passive muscle is compliant; that is, it can be easily stretched. The active muscle is stiff: One must apply great force to stretch it. The greater the muscle tension, the greater its stiffness. Superior athletes can develop high forces. The stiffness of their muscles, while active, exceeds the stiffness of their tendons (Figure 2.17). That is why elastic energy in elite athletes (for instance, during takeoffs) is stored primarily in tendons rather than in muscles. Tendon elasticity and a specific skill in using this elasticity in sport movement (takeoff, delivery) are important for elite athletes.

Neural mechanisms. Consider the neural mechanisms governing reversible muscle action during a drop jump landing. After the foot strike, there is a rapid change in both the muscle length and the forces developed. The muscles are forcibly stretched, and at the same time, muscle tension rises sharply. These changes are controlled and partially counterbalanced by the concerted action of two motor reflexes: *myotatic* (or *stretch*) *reflex* and *Golgi tendon reflex.*

These reflexes constitute two feedback systems that operate

- to keep the muscle close to a preset length (myotatic reflex; *length feedback*) and
- to prevent unusually high and potentially damaging muscle tension (Golgi tendon reflex; *force feedback*).

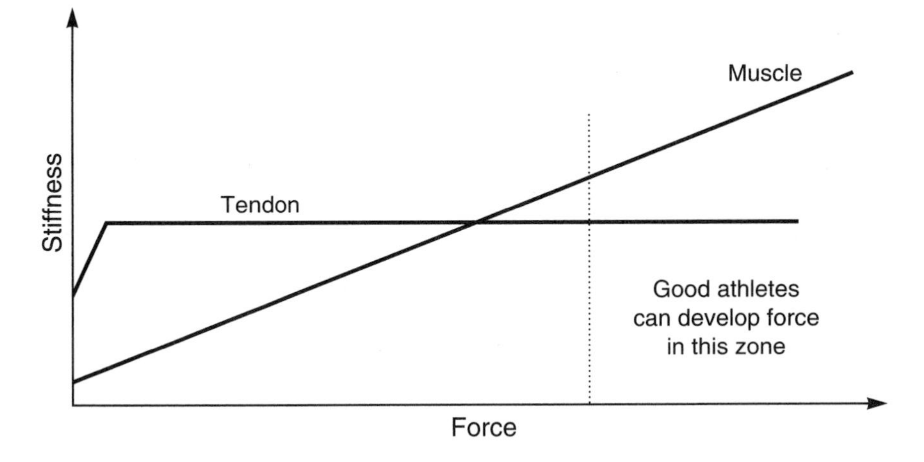

Figure 2.17 Stiffness of a muscle and a tendon. Since elite athletes develop high forces, the stiffness of the muscle, while active, exceeds tendon stiffness. The tendons are deformed to a greater extent than the muscles and thus store more elastic energy.

The myotatic reflex receptors (*muscle spindles*) are arranged parallel to muscle fibers that constitute the bulk of the muscle. When the muscle is stretched by an external force, the muscle spindles are also subjected to stretching. The stretching induces an increase in muscle spindle discharge. The latter causes an increased discharge of alpha-motoneurons and in turn a reflex contraction of the stretched muscle. This reflex contraction causes the muscle to return to its initial length in spite of the load changes applied to the muscle (*length feedback*).

Golgi tendon organs are arranged in series with the muscle fibers. These receptors are sensitive to forces developed in the muscle, rather than to length changes as with the muscle spindles. If muscle tension increases sharply, the Golgi tendon reflex evokes the inhibition of muscle action. The ensuing drop in muscle tension prevents the muscle and tendon from incurring damage (*force feedback*).

The efferent discharge to the muscle during the stretching phase of a stretch-shortening cycle is modified by the combined effects of the two reflexes mentioned earlier: the positive (excitatory) effect from the myotatic reflex and the negative (inhibitory) effect from the Golgi tendon reflex. During landing, a stretch applied to a leg extensor produces (via myotatic reflex) a contraction in that muscle; simultaneously, a high muscle tension sets up a Golgi tendon reflex in the same muscle, thus inhibiting its activity (Figure 2.18). If athletes, even strong ones, are not accustomed to such exercises, the activity of the extensor muscles during takeoff is inhibited by the Golgi tendon reflex. Because of this, even world-class weight lifters cannot compete with triple jumpers in drop jumping. As a result of specific training, the Golgi tendon reflex is inhibited and the athlete sustains very high landing forces without a decrease in exerted muscular force. The dropping height may then be increased.

■ *Muscles and Tendons as Springs in Series*

To visualize a stretch-shortening cycle, imagine two springs connected in series. The first spring (tendon) possesses given characteristics (stiffness, compliance) that do not change during motion. The characteristics of the second spring (muscle) vary and depend on the level of muscle activation.

When the muscle is relaxed, it is very compliant. If an external force is applied to such a muscle-tendon complex, the muscle can easily be stretched. The resistance to deformation is small, and only the muscle, not the tendon, is extended. However, if the muscle is activated, its resistance to the external pulling force increases. In this instance, the tendon rather than the muscle is deformed when a tensile force is applied.

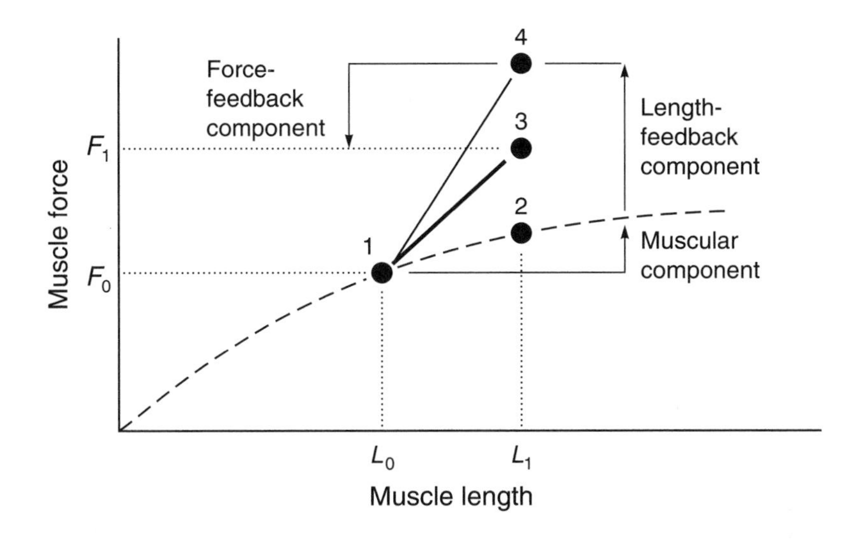

Figure 2.18 Mechanisms of enhanced force output in the stretch-shortening cycle. As a result of a stretch from L_0 to L_1, the muscle force increases from F_0 to F_1. Three functional components are responsible for the strength enhancement. (1) The muscular component—the force during lengthening increases due to muscle and tendon elasticity (stiffness). (2) The force output increases due to the length-feedback component—this component arises from the facilitatory spindle discharge (myotatic reflex). (3) The force-feedback component originating from Golgi tendon organs. The length-feedback component increases stiffness whereas the force-feedback component decreases it. The final outcome is the line from 1 to 3. The slope of this line defines the stiffness. The theory was originally developed by J.C. Houk and published in "Feedback Control of Muscle: A Synthesis of the Peripheral Mechanisms" 1974, in W.B. Mountcastle (Ed.), *Medical Physiology*, 13th ed. (pp. 668-677), St. Louis: Mosby. Note. From "Training of Muscle Strength and Power: Interaction of Neuromotoric, Hypertrophic and Mechanical Factors" by P.V. Komi, 1986, *International Journal of Sport Medicine*, **7** (Suppl.), p. 10. Reprinted by permission from the author.

The level of muscle activation is not constant, however, even when an athlete is trying to generate a maximal muscular effort. In addition to voluntary control, the muscles are under subconscious reflex control that is presumably realized on the spinal level. At least two reflexes are acting concurrently. One (stretch) reflex takes charge of maintaining the set muscle length—if the muscle is extended, it is additionally activated to resist the deformation force and to restore the original length. The second (Golgi organ) reflex prevents the muscle from injury due to excessive force—when the muscle tension or its rate is too high, the neural impulsation to the muscle from the spinal cord is inhibited.

The real intensity of muscle activation is a trade-off between the two reflexes (plus volitional muscle activation). The intensity of each reflex, which is not constant, determines the final outcome. When athletes are accustomed to sharp, forcible muscle-tendon stretching, for instance in drop jumps, the Golgi organ reflex is inhibited and high forces can be generated. The objective of drop jumping drills is, in this case, to accommodate the athletes to fast muscle stretching rather than to immediately generate large forces.

Since reversible muscle action is an element of many sport movements, it must be specifically learned or trained. Before 1960 such training was accidental, and improvement in this skill was a by-product of other exercises. Only since that time have exercises with reversible muscle action, such as drop jumps, been incorporated into training. Note that this training method has been erroneously called *pliometrics* by some. The term is not appropriate in this case, since reversible, not eccentric, muscle action is the training objective.

In beginners, performance in exercises with reversible muscle action can be improved through other exercises such as heavy weight lifting. In qualified athletes, this skill is very specific. Performances in drop jumps, for example, are not improved as a result of the usual strength exercising, even with heavy weights (Figure 2.19). Maximal muscular strength (F_{mm}) and forces produced in fast reversible muscle action (F_m) are not correlated in good athletes and should be treated, and trained, as separate motor abilities.

Posture, Strength Curves

The strength an athlete can develop in a given motion depends on body position (joint angles). For instance, the force F_m that one can apply to a barbell while lifting it from the floor (during a clean lift) depends on the height of the bar. The maximal force F_{mm} is exerted when the bar is near knee height (Figure 2.20).

■ *Why Do Elite Weight Lifters Start a Barbell Lift From the Floor Slowly?*

A good weight lifter imparts the greatest effort to a barbell, trying to accelerate it maximally, when the bar is approximately at knee-joint height. There are two reasons for this. First, at exactly this position the highest forces can be generated (Figure 2.20). And second (see discussion on velocity, p. 37), the force decreases when the

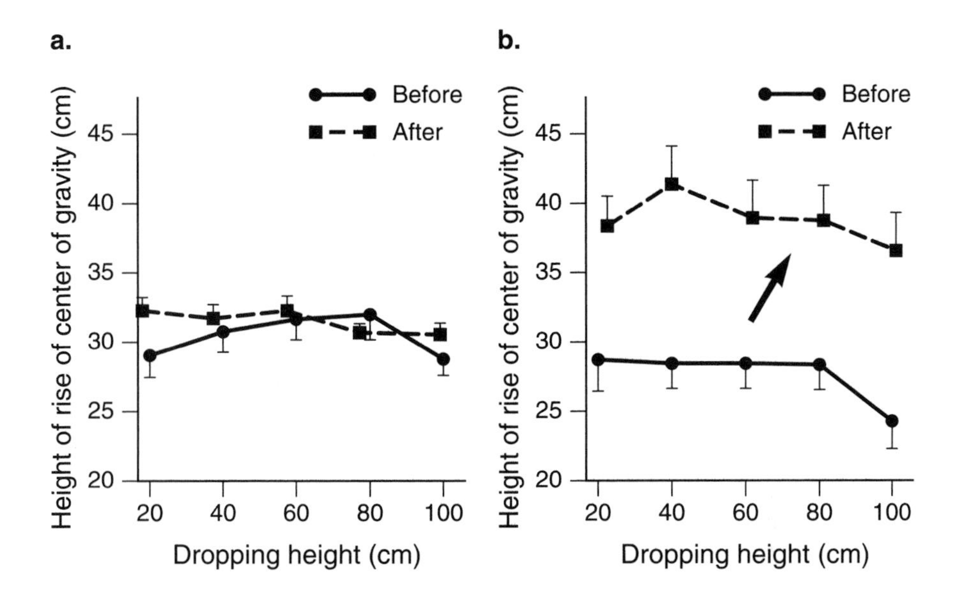

Figure 2.19 Changes in drop jump performances by experienced athletes after 24 weeks of training with: (a) heavy weights and (b) specific jumping training. (a) Heavy resistance (70-100% of F_m) training (n = 11). (b) Explosive (power) strength training (n = 10). Note. Adapted from "Changes in Electrical and Mechanical Behavior of Leg Extensor Muscles During Heavy Resistance Strength Training" by K. Håkkinen and P.V. Komi, 1985, *Scandinavian Journal of Sport Sciences*, **7**, pp. 55-64, and "Effect of Explosive Type Strength Training on Electromyographic and Force Production Characteristics of Leg Extensor Muscles During Concentric and Various Stretch-Shortening Cycle Exercise" by K. Håkkinen and P.V. Komi, 1985, *Scandinavian Journal of Sport Sciences*, **7** pp. 65-75. Reprinted by permission from the authors.

movement velocity increases (parametric force-velocity relationship). The barbell must approach the most favored body position (for force generation) at a relatively low velocity to impart maximal force to the bar. This two-phase technique is used by all elite weight lifters except in the light weight categories, by most 52-kg lifters, and by some at 56 kg. These athletes are short (below 150 cm), and the bar is located at knee-joint level in the starting position, before the lift.

This is an example of how two extrinsic factors of force generation (force posture and force velocity) are combined to develop maximal force values.

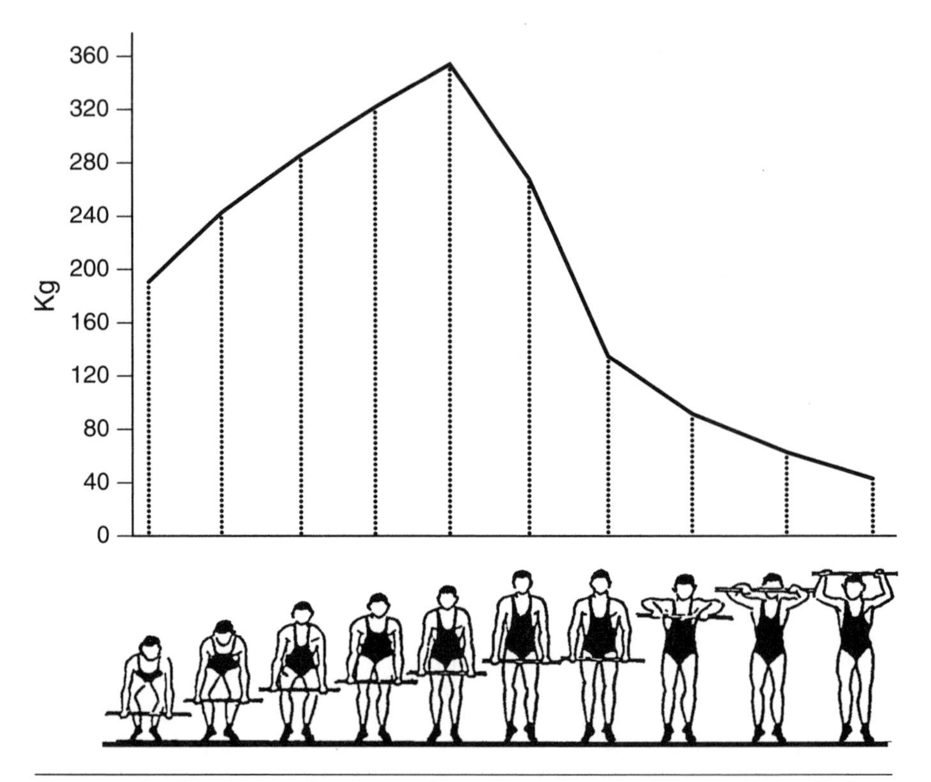

Figure 2.20 The maximal isometric force F_m applied to a bar at different body positions (at different heights of the bar). This is an example of the strength curve in a multijoint movement. Note. From *Biomechanics* (p. 203) by D.D. Donskoj and V.M. Zatsiorsky, 1979, Moscow: Fizkultura i Sport. Reprinted by permission from the publishing house.

The plot of the external force exerted by an athlete (or the moment of force) versus an appropriate measure of the body position (i.e., joint angle) is termed a *strength curve*. Strength curves assume three general forms: ascending, descending, and concave (Figure 2.21; see also examples in Figure 2.23).

The main factors determining these relationships are changes in *muscle length* and *muscle force arm*. The moments due to passive forces and antagonistic muscle activity can also contribute to the resultant joint moment (force) at extreme regions within the range of joint motion.

Muscle length varies with posture change. In turn, the tension that is produced by a muscle (subjected to a given amount of stimulation) depends on its length at the moment of measurement.

Muscle force changes with alterations of its length for two reasons. First, the area of overlapping actin and myosin filaments is changed, thus

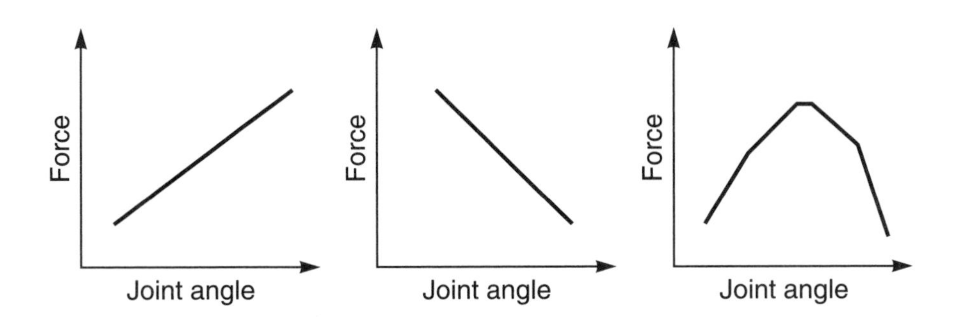

Figure 2.21 Three main forms of joint strength curves. Note. Adapted from "Mechanical Basis of Strength Expression" by J.G. Hay, 1992, in P.V. Komi (Ed.), *Strength and Power in Sport* (pp. 197-207), Oxford: Blackwell Scientific. Copyright 1992 by the International Olympic Committee. Adapted by permission from the publisher.

modifying the number of cross-bridge attachments that can be established (see later discussion of muscle dimensions). Second, the contribution of elastic forces, especially from parallel elastic components, is changed. Because of the interplay of these two factors, the relationship between the instantaneous muscle length and force production is complex. This relationship also varies with different muscle groups.

We can, however, disregard such complexity and accept as a general rule that *human muscles develop less force when they are shortened*. In contrast, higher forces are exerted by stretched muscles. Here is an example of how the total force of the ankle plantar flexors (triceps surae) can change at various ankle angles:

Angle, degrees	*Force, N*
140 (plantar flexion)	3,840
102 (plantar flexion)	4,630
90 (normal position)	5,600
78 (dorsiflexion)	5,950

Muscle force arms (i.e., distances from the axes of joint rotation and the lines of muscle action) are altered with changing joint angles. For instance, fourfold differences have been measured in the moment arm of the biceps brachii (long head) in assorted elbow angle positions; the force arm was 11.5 mm at the 180° angle (full extension) and 45.5 mm at the 90° angle of elbow flexion. Thus, if muscle tension was identical in each case, the moment of force developed by the muscle in elbow flexion would change fourfold. The external force (strength) would also be four times higher.

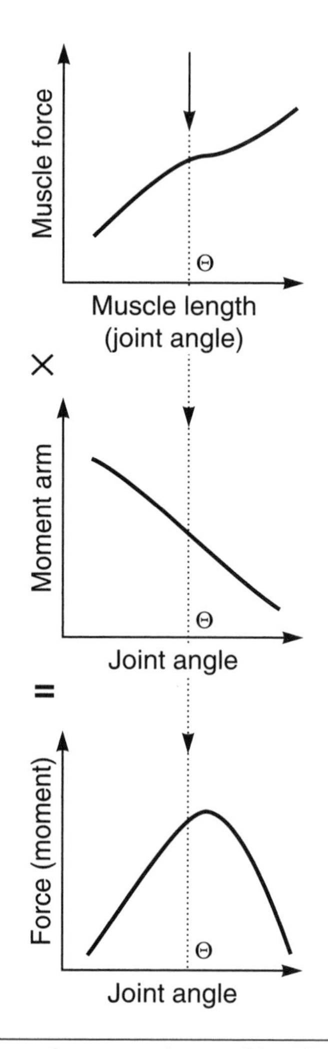

Figure 2.22 External muscular torque (strength) registered in joint angle Θ is the product of muscle tension and muscle lever arm length in this joint configuration.

The external torque generated by a muscle is a product of the force generated by the muscle and the muscle lever arm (Figure 2.22). Thus in order to estimate the external force, or moment, from the muscle at a given joint angle Θ (indicated by arrows in Figure 2.22), the force developed by the muscle at this angle should be multiplied by the muscle lever arm length at the corresponding angle. The combination of the factors shown leads to the existence of an angle-strength relationship for any one-joint movement (see Figure 2.23, a-c).

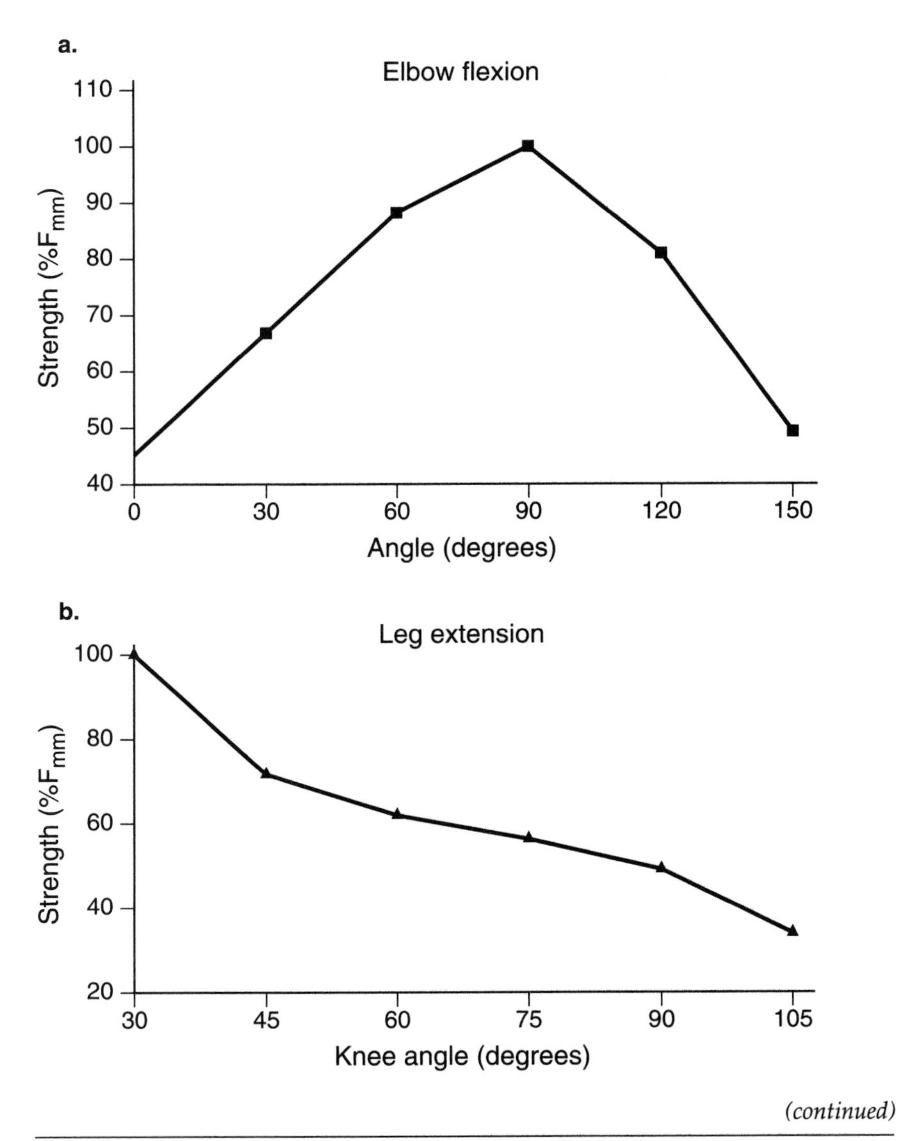

(continued)

Figure 2.23 Relationships between joint angle and isometric force for some joints and movements. The angles were defined from the anatomical position. Average data of 24 athletes. The elbow flexion force was measured with the forearm in a supinated position. The bilateral leg extension force was measured in a supine position with the angle of push 15° to the horizontal. The shoulder flexion measurements were made with subjects in a supine posture. The forearm was in a midrange position (between supinated and pronated). At – 30°, the arm was positioned behind the trunk. Note. Data from *Force-Posture Relationships in Athletic Movements* by V.M. Zatsiorsky and L.M. Raitsin, 1973, technical report, Moscow: Central Institute of Physical Culture.

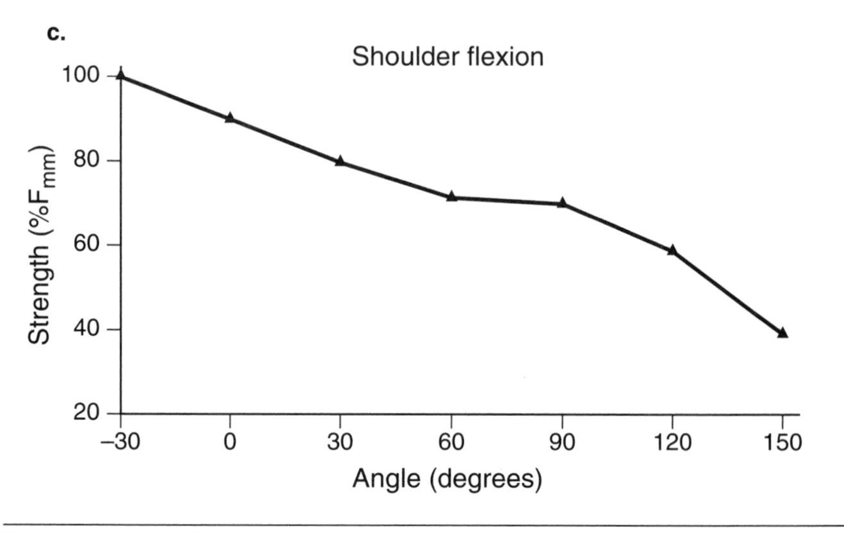

Figure 2.23 *(continued)*

When two-joint muscles are involved, as is usually the case, the muscular strength registered in one joint depends on the position or movement in adjacent joints. For instance, the force in knee flexion and extension is affected by the hip-joint position. In a sitting posture, the rectus femoris (knee extensor and hip flexor) is shortened and less knee extension force is developed than when the hip is in an upright position. The opposite is true for knee flexors (several of these are hip extensors also); in this case, strength is greater in the sitting position than in the standing position.

Even small changes in posture may periodically lead to significant strength gains or losses. For instance, changes of forearm position from pronation to supination increase elbow flexion strength by nearly one third.

For each motion, there are certain positions in which maximum strength values are achieved. In elbow flexion the optimal angle is 90°; in elbow and knee extension, it is 120°; and in hip extension, it is 155°. Coaches and athletes should be aware of the positions and angles in their main sport movement that are advantageous for maximum force production.

Strength values at the weakest positions (the so-called "sticking points") are also very important. The heaviest weight that is lifted through a full range of joint motion cannot be greater than the strength at the weakest point. This weight taxes only a certain percentage of the maximum force at other joint angles. Many consider this a disadvantage of exercises using free weights. Special cams of variable radii are used in some strength training devices to provide maximal or at least near-maximal muscular tension throughout the full range of joint motion. This is achieved by changing the moment arm of the machine so that the resistance, in turn, varies. However, the development of the maximal force throughout the full range of joint motion is not natural for human movements, including sport movements. Therefore, many coaches and elite athletes are against the use of such devices.

Three different approaches have been developed to maximize the training benefit of resistance work. In the first, maximal force is developed in the weakest body position (termed the *peak-contraction principle*). In the second approach, near-maximal force is attained throughout the complete range of joint motion (*accommodating resistance*). In the third, exercises are constructed to develop maximal force at the precise angular position in which maximal efforts are developed during the main sport movement. This third approach is called the *accentuation* of muscular efforts. These concepts will be discussed in chapter 6.

Summary

An athlete can perform a given motor task, such as throwing, lifting, or jumping, with different levels of effort. When effort is maximal, the athlete attains a *maximal muscular performance* for the given task. Each motor task is characterized by certain variables called parameters—such as resistance, the angle of slope in uphill running, or the weight of the object—and the magnitudes of these parameters.

If the parameters of a motor task are changed systematically, the *parametric relationship* between the dependent variables of maximal muscular performance can be established. The parametric relationship between the maximal force (F_m) and maximal velocity (V_m) is negative: The higher the force, the lower the velocity. The highest maximal force (F_{mm}) is termed the maximum maximorum force. The dependence between F_{mm} values and the maximal velocity (V_m) at a given parameter proportion is called the *nonparametric relationship*, a correlation that is typically positive (i.e., the higher the force, the greater the velocity). The strength of the correlation depends on the parameter values: The greater the resistance, the higher the coefficient of the correlation.

Muscular strength is the ability to produce maximum maximorum external force F_{mm}. It can be generated and measured only at certain parameter values of a motor task, such as muscular force exerted on a heavy implement. When athletes attempt to produce maximal force, the generated force values depend on the motor task. Even when the "geometry" of a motion (e.g., involved body limbs, movement trajectory) is fixed, the resulting force varies.

Several factors determine differences in force values across motor tasks. These factors are classified as extrinsic (external) and intrinsic (internal). The force exerted by an athlete on an external body depends not only on the athlete but also on external factors, in particular the type of resistance (such as elasticity, inertia, gravity force, hydrodynamic force).

The *type of resistance* influences the pattern of the force produced. Imagine that the same arm motion (e.g., in a lateral-medial direction) is performed against different resistance: first, springs, and then, viscosity (the arm moves in tough dough). In the first instance, the resistance increases in proportion to the movement amplitude; in the second, resistance is proportional to the movement velocity. Often the resistance provided by a strength exercise apparatus does not resemble the type of resistance found in natural sport movements. This is detrimental to the efficiency of strength training.

Several intrinsic characteristics of motor tasks are important for producing the maximal force. *Time available for force development* is a crucial factor in many sport events. The time required to produce maximal force is typically longer than the time available for the manifestation of strength in real sport movements. Thus the rate of force development, rather than the absolute force itself, is the crucial factor in a successful athletic performance. The relative contributions of the maximal force and the rate of force development depend on the level of athletic performance. The higher the performance, the shorter the time available for force production and thus the greater the importance of the rate of force development. The ability to produce maximal forces in minimal time is called *explosive strength*. Strong people do not necessarily possess explosive strength.

Movement velocity influences the magnitude of the force that can be produced; the higher the velocity, the smaller the force (parametric relationship). Thus the lower the movement velocity and, consequently, the greater the force values produced during the natural athletic movement, the greater the contribution of F_{mm} (and also of heavy resistance training) toward athletic performance.

Direction of movement (i.e., whether the muscle is shortening or lengthening during a motion) is a matter of primary importance. The highest forces are generated during eccentric muscle action as well as during reversible muscle action, when the muscle is forcibly stretched and then permitted to shorten. Such a stretch-shortening cycle is an innate part of many athletic movements. The magnitude of the force produced during the stretch-shortening muscle action, as well as the magnitude of the stored and recoiled potential energy of deformation, depends upon both the elastic properties of muscles and tendons and the neural control of muscle activity. The interplay of two spinal reflexes (stretch reflex and Golgi organ reflex) is considered to be a major factor toward determining neural inflow to the muscle during the stretch-shortening cycle.

Furthermore, the magnitude of the manifested muscle force depends largely upon *body posture*. For one-joint motions, *strength curves* are calculated (i.e., the force-angle relationships). Two main factors affecting the force-angle curve are changes in muscle tension forces and changes in force arms. In multijoint body movements, the strongest as well as the weakest (sticking) points exist throughout the whole range of motion at which maximal (minimal) force values are manifested.

CHAPTER 3

Athlete-Specific Strength

In the last chapter we looked at how strength depends on various factors specific to the tasks within a given sport or activity. We turn now to the factors that affect maximal forces produced by individual athletes, and how they may vary between people, that is, the determining factors in a comparison across athletes. We conclude the chapter and examination of the determinants of strength with a taxonomy to help you consolidate and sort what you have learned in chapters 2 and 3.

Individual athletes generate different maximal forces when they perform similar motions. These variations stem mainly from two factors:

- the maximal force capabilities of individual muscles (so-called *peripheral factors*),

- the coordination of muscle activity by the central nervous system (*central factors*). Two aspects of neural coordination are discernible: *intramuscular* coordination and *intermuscular* coordination.

This is not a book on physiology, so we will look only briefly at these factors to clarify what is most relevant to strength training.

Muscle Force Potential (Peripheral) Factors

Among the peripheral factors affecting muscle force potential, muscle dimensions seem to be the most important. Muscle mass and dimensions are affected by training, of course, and by other factors, including nutrition and hormonal status.

Muscle Dimensions

It is well known that muscles with a large physiological cross-sectional area produce higher forces than similar muscles with a smaller cross section. This is true regardless of muscle length. With heavy resistance training in which the muscle cross section is increased, there is typically an accompanying increase in maximal strength.

Skeletal muscle consists of numerous *fibers*, or long, cylindrical muscle cells. Each fiber is made up of many parallel *myofibrils*, which consist of longitudinally repeated units called *sarcomeres*. Sarcomeres in turn include thin *filaments* consisting largely of the protein *actin* and thick filaments of the protein *myosin*. The actin and myosin filaments partially overlap. Myosin filaments have small outward helical projections called *cross bridges*. These cross bridges end with *myosin heads* that make contact (*cross-bridge attachments* or *links*) with the thin filaments during contraction. According to the *sliding-filament theory*, shortening of the sarcomere, and hence the muscle fiber, occurs as a result of the active relative sliding of the actin filaments between the myosin filaments.

The force produced by a muscle is the outcome of activity of muscle subunits (sarcomeres, myofibrils, muscle fibers). The maximal force produced by a *sarcomere* depends to some extent on the total number of myosin heads available for the cross-bridge links with actin filaments. The total number of cross-bridge links in a given sarcomere is apparently the product of

- the number of actin and myosin filaments, i.e., the cross-sectional area of all the filaments, and
- the number of myosin heads that can interact with actin filaments, i.e., sarcomere length.

Muscles with long sarcomeres (longer actin and myosin filaments) exert greater force per unit of cross-sectional area because of the greater extent of possible overlap.

All the sarcomeres of one *myofibril* work in series. The force exerted by, or on, any element of a linear series (i.e., by any sarcomere in the myofibril) is equal to the force developed in each of the other elements in the series. Therefore, all sarcomeres of the myofibril exert the same force, and the force registered at the ends of the myofibril does not depend on its length.

The force produced by a muscle *fiber* is limited by the number of actin and myosin filaments and consequently by the number of myofibrils working in parallel. The differences in parallel and serial action of sarcomeres are listed in Figure 3.1 for the example of two "fibers" consisting of two sarcomeres each. To estimate the muscle potential in force production, instead of calculating the number of filaments, researchers determine their total cross-sectional area. The ratio of the filament area to the muscle fiber area is called *filament area density*.

Strength exercise can increase the number of myofibrils per muscle fiber and filamental area density; thus there is a rise in both muscle cell size and strength. We know little about the influence of strength training on sarcomere length.

The capacity of a *muscle* to produce force depends on its physiological cross-sectional area, and particularly on the number of muscle fibers in the muscle and the cross-sectional areas of the fibers.

It is commonly known that the size of a muscle increases when it is subjected to a strength training regimen. This increase is called *muscle hypertrophy* and is typically displayed by bodybuilders. Muscle hypertrophy is caused by

- an increased number of motor fibers (this is called *fiber hyperplasia*) or
- the enlargement of cross-sectional areas of individual fibers (*fiber hypertrophy*).

Recent investigators have found that both hyperplasia and hypertrophy contribute to muscle size increase. However, the contribution of fiber hyperplasia is rather small and may be disregarded for practical purposes of strength training. Muscle size increases are caused mainly by individual fiber size increases, not by the gain in fibers (through fiber splitting). People with large numbers of small (thin) muscle fibers have a greater potential to become good weight lifters or bodybuilders than do people with small numbers of fibers in their muscles. The size of individual fibers, and consequently the size of the muscles, increases as a result of training. The number of fibers is not changed substantially.

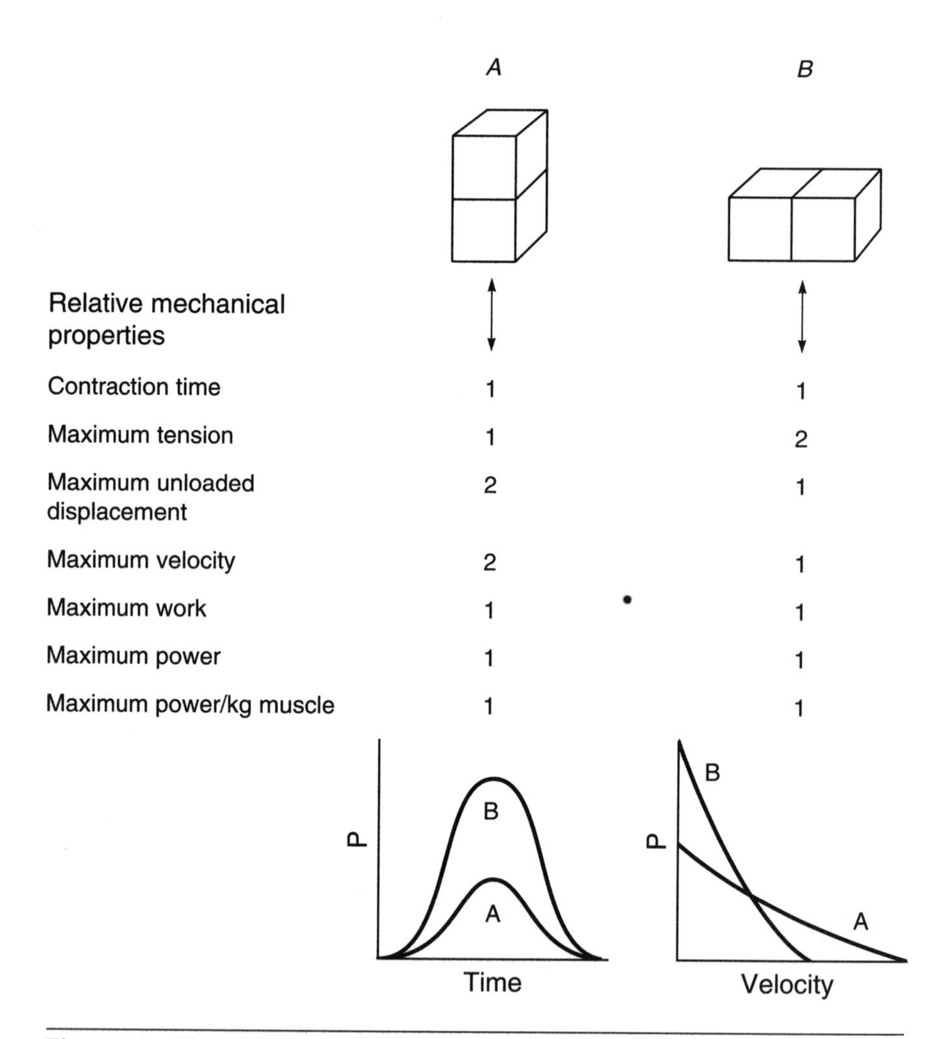

Figure 3.1 The relative effects of different arrangements of sarcomeres, in series and parallel, on the mechanical properties of a muscle fiber. In addition,the relative isometric and isotonic properties are illustrated at the bottom of the figure for condition A and condition B. Note. From "Morphological Basis of Skeletal Muscle Power Output" by V.R. Edgerton, R.R. Roy, R.J. Gregor, and S. Rugg, 1986, in N.L. Jones, N. McCartney, and A.J. McComas (Eds.), *Human Muscle Power* (p. 44), Champaign, IL: Human Kinetics. Copyright 1986 by Human Kinetics. Reprinted by permission.

Two types of muscle fiber hypertrophy can be schematically depicted: *sarcoplasmic* and *myofibrillar* (Figure 3.2).

Sarcoplasmic hypertrophy of muscle fibers is characterized by the growth of sarcoplasm (semifluid interfibrillar substance) and noncontractile proteins that do not directly contribute to the production of muscle force. Specifically, filament area density in the muscle fibers decreases, while the cross-sectional area of the muscle fibers increases, without an accompanying increase in muscle strength. *Myofibrillar hypertrophy* is an enlargement of the muscle fiber as it gains more myofibrils and, correspondingly, more actin and myosin filaments. At the same time, contractile proteins are synthesized and filament density increases. This type of fiber hypertrophy leads to increased muscle force production.

Heavy resistance exercise can lead to both sarcoplasmic and myofibrillar hypertrophy of muscle fibers. However, depending on the training routine, these types of fiber hypertrophy are manifested to varying degrees. Myofibrillar hypertrophy is typically found in elite weight lifters (if the training program is designed properly), whereas sarcoplasmic hypertrophy is typically seen in bodybuilders. Except for very special cases in which the aim of heavy resistance training is to achieve body weight gains, athletes are interested in inducing myofibrillar hypertrophy. Training must be organized to stimulate the synthesis of contractile proteins and to increase filament muscle density.

A common belief in strength training is that exercise activates protein catabolism (breakdown of muscle proteins), creating conditions for the enhanced synthesis of contractile proteins during the rest period (*break down and build up theory*). During strength exercises, muscle proteins are forcefully converted into more simple substances ("breaking down"); during restitution (anabolic phase), the synthesis of muscle proteins is vitalized. Fiber hypertrophy is considered to be a *supercompensation* of muscle proteins.

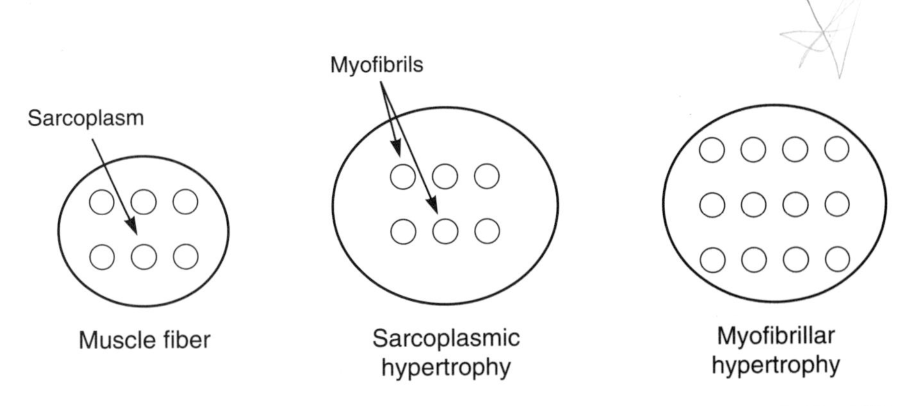

Figure 3.2 Sarcoplasmic and myofibrillar hypertrophy.

The mechanisms involved in muscle protein synthesis, including the initial stimuli triggering the increased synthesis of contractile proteins, have not been well established at this time. A few hypotheses that were popular among coaches 20 to 30 years ago are now completely disregarded, including these:

- The *blood overcirculation* hypothesis suggests that increased blood circulation in working muscles is the triggering stimulus for muscle growth. One of the most popular methods of bodybuilding training, called *flushing* (see chapter 7, muscle mass), is based on this assumption. It has been shown, however, that active muscle hyperemization (i.e., the increase in the quantity of blood flowing through a muscle) caused by physical therapeutical means does not, in itself, lead to the activation of protein synthesis.

- The *muscle hypoxia* hypothesis, in contrast to the blood overcirculation theory, stipulates that a deficiency, not an abundance, of blood and oxygen in muscle tissue during strength exercise triggers protein synthesis. Muscle arterioles and capillaries are compressed during resistive exercise and the blood supply to an active muscle is restricted. Blood is not conveyed to muscle tissue if the tension exceeds approximately 60% of maximal muscle force.

  However, by inducing a hypoxic state in muscles in different ways, researchers have shown that oxygen shortage does not stimulate an increase in muscle size. Professional pearl divers, synchronized swimmers, and others who regularly perform low-intensity movements in oxygen-deficient conditions do not have hypertrophied muscles.

- The *adenosine triphosphate (ATP) debt theory* is based on the assumption that ATP concentration is decreased after heavy resistive exercise (about 15 repetitions in 20 s per set were recommended for training). However, recent findings indicate that, even in a completely exhausted muscle, the ATP level is not changed.

A fourth theory, although it has not been validated in detail, appears more realistic and appropriate for practical training—the *energetic theory* of muscle hypertrophy. According to this hypothesis, the crucial factor for increasing protein catabolism is a shortage in the muscle cell of energy available for protein synthesis during heavy strength exercise. The synthesis of muscle proteins requires a substantial amount of energy. The synthesis of one peptide bond, for instance, requires energy liberated during the hydrolysis of two ATP molecules. For each instant in time, only a given amount of energy is available in a muscle cell. This energy is spent for the anabolism of muscle proteins and for muscular work. Normally, the amount of energy available in a muscle cell satisfies these two requirements. During heavy resistive exercise, however, almost all available energy is conveyed to the contractile muscle elements and spent on muscular work (Figure 3.3).

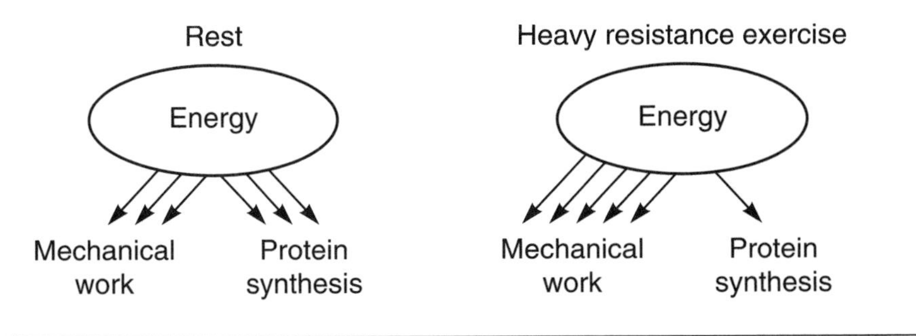

Figure 3.3 Energy supply at rest and during light and heavy resistance exercise.

Since the energy supply for the synthesis of proteins decreases, protein degradation increases. The uptake of amino acids from the blood into muscles is depressed during exercise. The mass of proteins catabolized during heavy resistive exercise exceeds the mass of protein that is newly synthesized. As a result, the amount of muscle proteins decreases somewhat after a strength workout, while the amount of protein catabolites (e.g., the concentration of nonprotein nitrogen in the blood) rises above its resting value. Then, between training sessions, protein synthesis is increased. The uptake of amino acids from the blood into muscles is above resting values. This repeated process of enhanced degradation and synthesis of contractile proteins may result in the supercompensation of protein (Figure 3.4). This principle is similar to the overcompensation of muscle glycogen that occurs in response to endurance training.

Whatever the mechanism for stimulating muscle hypertrophy, the vital parameters of a training routine that induce such results are exercise intensity (the exerted muscular force) and exercise volume (the total number of repetitions, performed mechanical work). The practical aspects of this theory will be described in chapter 4.

Body Weight

Muscle mass constitutes a substantial part of the human body mass or body weight. (In elite weight lifters, muscle mass is about 50% of body weight). That is why, among equally trained individuals, those with greater body weight demonstrate greater strength.

The dependence of strength on weight is seen more clearly when tested subjects have equally superb athletic qualifications. World record holders in weight lifting have shown a very strong correlation between performance level and body weight, 0.93. The correlation for participants at the world championships has been 0.80; and among those not involved in sport activities, the correlation has been low and may even equal zero.

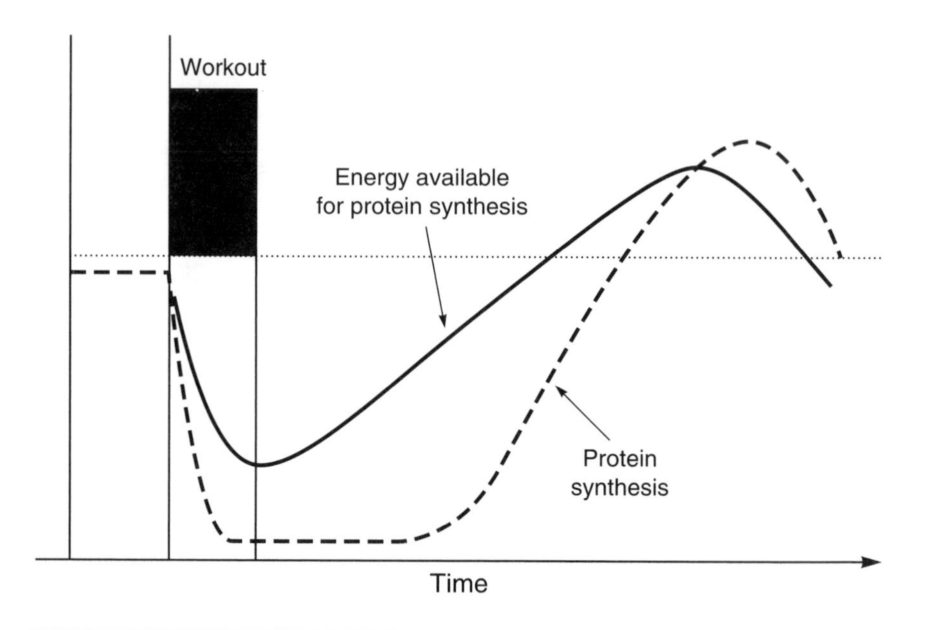

Figure 3.4 Energetic potential of a muscle cell and the rate of protein anabolism. Note. Adapted from "Influence of Exercise on Protein Metabolism" by A.A.Viru, 1990, in A.A. Viru (Ed.), *Lectures in Exercise Physiology* (pp. 123-146), Tartu, Estonia: The Tartu University Press. Adapted by permission from the author.

To compare the strength of different people, the strength per kilogram of body weight (termed *relative strength*) is usually calculated. On the other hand, muscular strength, when not related to body weight, is called *absolute strength*. Thus, the following equation is valid:

$$\text{Relative strength} = \text{Absolute strength}/\text{Body weight}$$

With an increase in body weight, among equally trained athletes of various weight classes, absolute strength increases and relative strength decreases (Figure 3.5).

For instance, leading international athletes in the 60-kg weight category lift a barbell in the clean and jerk that is heavier than 180 kg (the world record equals 190.0 kg). Their relative strength in this exercise exceeds 3.0 (180 kg of force/60 kg of body weight = 3.0). The body weight of athletes in the super heavy weight division, on the other hand, must be above 110 kg and is typically 130 to 140 kg. If the best athletes of this weight class had a relative strength of 3.0 kg of force per kilogram of body weight, they would lift approximately 400 kg in the clean and jerk. In reality, the world record in this weight class is about 270 kg.

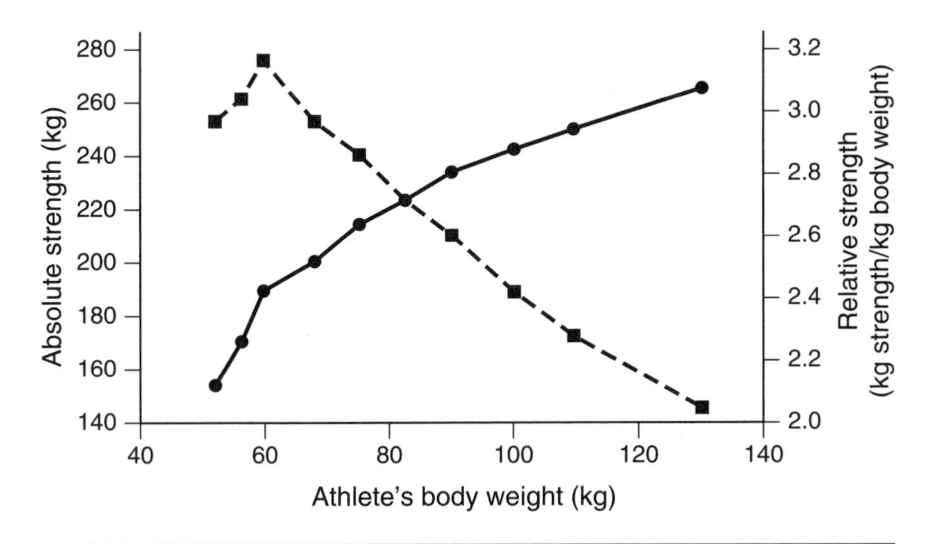

Figure 3.5 Absolute (solid line) and relative (dashed line) strengths of elite weight lifters at different weight classes. World records in clean and jerk lift (November 1991) serve as indices of absolute strength.

Because of their great relative strength, athletes of small body dimensions have an advantage in lifting their own bodies. Elite wrestlers of light weight classes can usually perform more than 30 pull-ups on a horizontal bar; for athletes in the super heavy weight category, 10 pull-ups is an excellent achievement.

■ *Why Do Athletes From Various Sports Have Different Body Dimensions?*

Why are gymnasts short? (The height of the best male gymnasts is usually in the range of 155 to 162 cm; female gymnasts are typically 135 to 150 cm tall and often even shorter.) Because they have to lift their own body and nothing else, relative, not absolute, strength is important in gymnastics. Short athletes have an advantage in this sport.

Why are the best shot-putters tall and heavy (but not obese)? Because here absolute strength is important. Athletes with large body dimensions have a distinct advantage in this sport.

To see what causes such discrepancies, imagine two athletes, A and B, with equal fitness levels but different body dimensions. One of them is 1.5

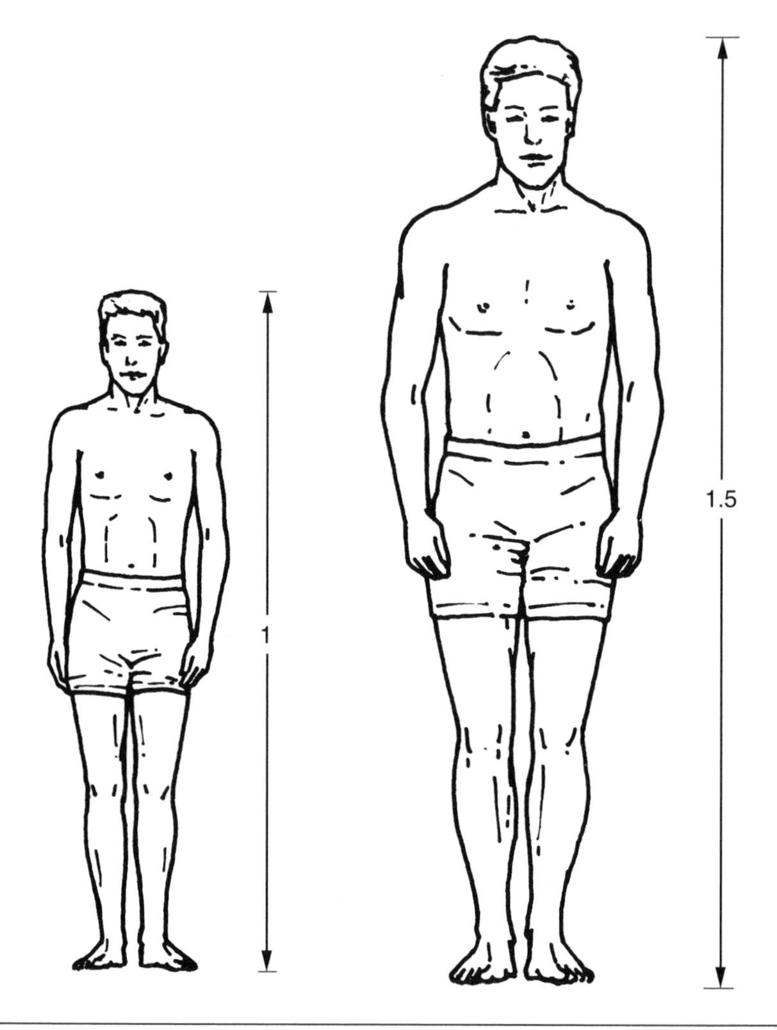

Figure 3.6 Two athletes of different body dimensions.

times as tall as the other (Figure 3.6). Their heights are 140 and 210 cm, and all anteroposterior and frontal diameters are also in the proportion 1:1.5.

Compare the linear measures (length of segments, diameters), surface measures (physiological cross-sectional area, body surface), and volume measures (volume, body mass):

Measure	*A*	*B*
Linear	1	1.5
Area (and strength)	1	$1.5^2 = 2.25$
Volume (and body weight)	1	$1.5^3 = 3.375$

Thus, the proportion for body height is 1:1.5; the proportion for area (including muscle physiological cross-sectional area) is 1:2.25; and the proportion for volume and weight is 1:3.375. Athlete B is 2.25 times stronger than athlete A, but also 3.375 times heavier. Athlete B has the advantage in absolute strength, and athlete A the advantage in relative strength.

The relationship between body weight and strength can then be analyzed using simple mathematics. Taking into account that

$$W = aL^3$$

where W is the body weight, L is the linear measure, and a is a constant, we can write

$$L = aW^{1/3}$$

Since strength (F) is proportional to muscle physiological cross-sectional area, it is also proportional to L^2:

$$F = aL^2 = a(W^{1/3})^2 = aW^{2/3} = aW^{0.666}$$

or, in logarithmic form,

$$\log_{10} F = \log_{10} a + 0.666 \log_{10} W.$$

We can validate the last equation by using, for instance, the world records in weight lifting. With this objective the logarithm of body weight is plotted in Figure 3.7 against the logarithm of weight lifted by an athlete. The regression coefficient is 0.646 (close to the predicted 0.666), proving that the equation is valid. Such an equation (or corresponding tables such as Table 3.1) can be used to compare the strength of people with different body weights. The table shows that a 100-kg force in the 67.5 kg weight class corresponds to 147 kg in super heavy weight lifters.

For linemen in football, super heavy weight lifters, and throwers, among others, absolute strength is of great value. For sports in which the athlete's body rather than an implement is moved, the relative strength is most important. Thus, in gymnastics, the "cross" is performed only by those athletes whose relative strength in this motion is near 1 kg per kilogram of body weight (Table 3.2). Because the gymnast does not suspend the entire body (there is no need to apply force to maintain

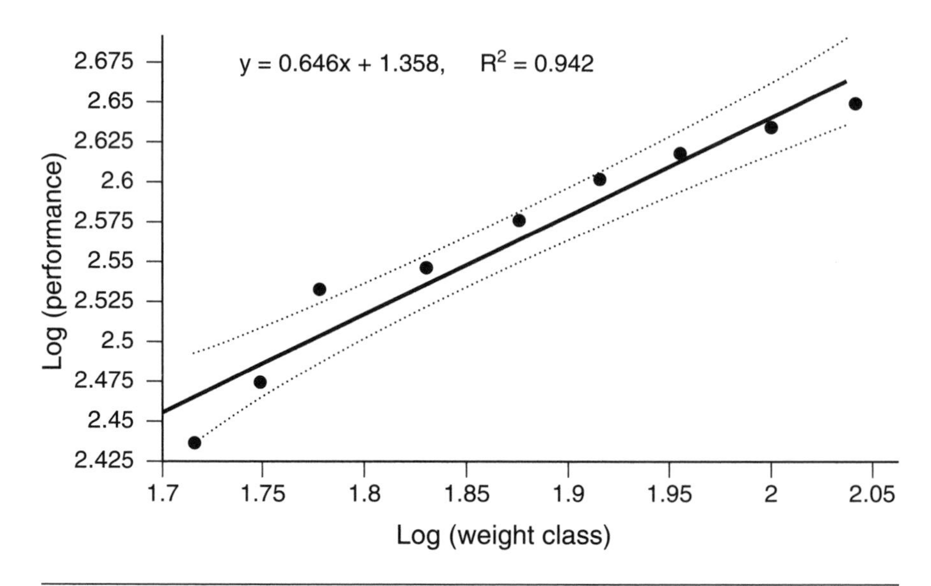

Figure 3.7 The relationship between athlete strength and body weight. The world records in weight lifting (snatch plus clean and jerk lifts) for athletes of different weight categories are used as indices of maximal strength. Because the body weight of athletes in the super heavy class (above 110 kg) is not precisely controlled by the rules, these data are not included in the analysis. The world records are for November 1991. A 90% confidence interval for the regression is shown. The world record in the 60-kg weight class (Suleymanoglu, 342.5 kg) is obviously above the average level of world records in other weight categories. Note logarithmic scale.

handholds), the cross can be performed when relative strength is slightly less than 1.0.

In sports in which absolute strength is the main requirement, athletes should train in a manner that stimulates lean muscle mass gain. As weight increases, the percentage of body fat must remain constant, or even decrease, to ensure that the weight gain is primarily in lean body mass.

An increase in relative strength may be accompanied by different changes in body weight. Sometimes it is accompanied by stabilization or even weight loss. Table 3.3 illustrates this phenomenon for an athlete who lost weight and increased her performance. Proper eating habits and regular weight control are necessary for all athletes. Weekly weigh-ins and regular determinations of body composition (skinfold measurements, underwater weighing) are an excellent idea.

A common athletic practice is to reduce body weight before competition. Athletes "make weight" in order to increase their relative strength

Table 3.1 Equivalent Strength Levels for Athletes of Different Weight Categories (kg)

Weight class, kg							
56	60	67.5	75	82.5	90	110	120
44	46	50	54	57	61	69	73
53	55	60	64	69	73	83	88
62	65	70	75	80	85	96	103
71	76	80	86	91	97	111	117
79	83	90	97	103	109	125	132
88	92	100	107	114	121	139	147
132	139	150	161	171	182	208	220
177	185	200	215	229	242	277	293
221	231	250	290	285	303	346	367
265	277	300	322	343	363	415	425

Note. Data are from "Applied Aspects of the Analysis of the Relationships Between the Strength and Body Weight of the Athletes" by V.M. Zatsiorsky and I.F. Petrov, 1964, *Theory and Practice of Physical Culture*, **27**(7), pp. 71-73.

Table 3.2 Maximal Force of Arm Adduction in "Cross" Position of Two World Champions in Gymnastics

Name	Arm adduction force, kg	Body weight, kg	Force excess over the weight, kg	Relative force, kg of force per 1 kg of body weight	"Crosses" in a composition
Azarian, A.	89	74	15	1.20	5–6
Shachlin, B.	69.2	70	−0.8	0.98	1–2

Note. Adapted from *Strength Testing of Elite Athletes* by A.A. Korobova and A.B. Plotkin, 1961, technical report #61-105 (p. 48), Moscow: All-Union Research Institute of Physical Culture.

Table 3.3 Body Weight Changes and Some Indirect Indices of Relative Strength in 1960 Olympic Games Champion (Long Jump) V. Krepkina

Age	Weight, kg	Height, m	Weight/ height	Standing jump, cm	Long jump, cm	Sprint 100 m, s
16	64	1.58	40.5	214	490	13.6
24	55	1.58	34.5	284	617	11.3

Note. Data are from *Motor Abilities of Athletes* (p. 26) by V.M. Zatsiorsky, 1966, Moscow: Fizkultura i Sport.

and improve performance. In sports with weight categories, such as wrestling and judo, athletes succumb to this practice to be eligible for a division lower than their usual weight division. Food restriction, fluid deprivation, and dehydration induced by thermal procedures such as the sauna are used toward this end.

This strategy is acceptable when properly employed (weight loss should not exceed 1 kg/week in "average" athletes and 2.5 kg in elite ones). However, extreme weight reduction is detrimental to athletic performance and is unsafe. Rapid loss of body weight over short periods of time leads to lean tissue and water loss rather than the loss of fat. In addition, there is a depletion of glycogen stores, the most important energy source for high-intensity performance. The athlete's capacity decreases either as a consequence of reduced carbohydrate availability or as a result of the effects of the disturbed fluid balance. Thus, it is important that athletes follow only long-term, planned weight reduction programs with food restriction in the range of 2 to 4 mJ/day (500–1,000 kcal/day) below normal energy intake.

Abuses associated with extreme and rapid weight loss, such as the use of rubber suits, laxatives, enemas, induced vomiting, and diuretics, cannot be justified. Diuretics, for instance, are considered doping; their use is prohibited by the Medical Commission of the International Olympic Committee (IOC). Unfortunately, in spite of all efforts to discourage the malpractice of rapid weight reduction, many athletes continue to lose weight through unacceptable and unsafe methods. Another caution is that all weight loss methods, even mild ones, are suitable only for mature athletes. They should be vigorously opposed for children and teenagers.

■ Gymnasts at Risk

Christy Henrich, one of the best American gymnasts of the 1980s, is a well-known example of the tragic consequences of eating problems. When she weighed 95 lbs., she was told by her coaches that she was too fat to make the Olympic team. She began a life of anorexia and bulimia, still missing the Olympics by a fraction of a point. Less than a decade later she died at age 22, weighing just 52 lbs. Coaches should comment about weight issues thoughtfully and carefully.

The alternative to weight loss is an increase in relative strength through gains in muscle mass. This is completely justified, and athletes should not be wary of muscular growth (for muscles carrying the main load in their sport movements).

■ Growth and Strength

As children and teenagers become taller and heavier, their relative strength should decrease. This often happens, especially during pubertal growth spurts. It is not uncommon for 8-year-old boys and girls to show comparatively high values of relative strength, for instance, to perform 10 or 12 chin-ups. But if they do not exercise regularly, they will not be able to repeat these achievements when they are 16.

Typically, however, the relative strength of children does not decrease during childhood and puberty, because during the maturation process the muscles of mature individuals produce a greater force per unit of body mass. Thus, two concurrent processes with opposite effects take place during childhood and puberty: growth (i.e., an increase in body dimensions) and maturation. Due to growth, relative strength decreases; at the same time, due to maturation, it increases. The superposition of these two processes determines the manifested strength advancement (or decline). The interplay of the two concurrent processes of child development is important in the preparation of young athletes.

Consider this example of how the best soviet/Russian male gymnasts train (for many years they have been the best in the world). They learn all the main technical stunts, including the most difficult ones, before the age of 12 or 13 when the puberty growth spurt begins. During the puberty period (13–16 years of age) they learn very few, if any, new technical elements. In training during this period they concentrate on conditioning, especially strength training and specific endurance training, and stability of performance. All compulsory and optional routines are trained (to achieve high stability of performance and gain specific endurance) rather than new elements and single stunts. Great attention is paid to strength development. As a result, at 17 to 18 years of age the gymnasts are prepared to compete at the international level. Dmitri Belozerchev won an all-around world championship when he was 16.

As the complexity of the optional routines increases, the most difficult stunts are performed (during training sessions) not by contemporary Olympic and world champions, but by their young counterparts (i.e., 12- and 13-year-old boys who are preparing at this time to compete at the 2000 and 2004 Olympic games).

Other Factors (Nutrition, Hormonal Status)

Strength training activates the synthesis of contractile muscle proteins and causes fiber hypertrophy only when there are sufficient substances for protein repair and growth. The building blocks of such proteins are amino acids, which must be available for resynthesis in the rest period after workouts.

Amino acids are the end-products of protein digestion (or hydrolysis). Some amino acids, termed *essential* or *indispensable,* cannot be produced by the body and must be provided by food. Amino acids supplied by food pass unchanged through the intestinal wall into the general blood circulation. From there they are absorbed by the muscles according to the specific amino acid needed by that muscle to build up its own protein. In practical terms, then,

- the full assortment of amino acids required for protein anabolism must be present in the blood during the restitution period; and
- proteins, especially essential ones, must be provided by the proper kinds of foods in sufficient amounts.

Athletes in sports such as weight lifting and shot putting, in which muscular strength is the dominant motor ability, need at least 2 g of protein per kilogram of body weight. In superior athletes during periods of *stress training,* when the training load is extremely high, the protein demand is up to 3 g per kilogram of body weight a day. This amount of protein must be provided by foods with a proper assortment of essential amino acids. It is important to note that the actual requirements are not for protein but rather for selected amino acids.

In addition to the amino acid supply, the hormonal status of an athlete plays a very important role. Several hormones secreted by different glands in the body affect skeletal muscle tissue. These effects are classified as either catabolic, leading to the breakdown of muscle proteins, or anabolic, leading to the synthesis of muscle proteins from amino acids. Among the anabolic hormones are testosterone, growth hormone (somatotropin), and somatomedins (insulin-like growth factor).

The concentrations of these hormones in the blood largely determine the metabolic state of muscle fibers. The serum level of testosterone is lower in females than in males, and therefore strength training does not elicit the same degree of muscle hypertrophy in females as in males. Strength training elicits changes in the level of anabolic hormones circulating in the blood. These changes may be acute (as a reaction to one workout) or cumulative (long-term changes in resting levels). For instance, strength training elicits increases in resting serum testosterone concentrations and induces an acute elevation in the level of circulating testosterone. A relatively high positive correlation ($r = 0.68$) has been found between the ratio of serum testosterone to sex hormone-binding globulin (SHBG) and concomitant gains in competitive weight-lifting results for the clean and jerk lift (Figure 3.8).

Serum somatotropin levels are significantly elevated during exercise with heavy weight (70–85% of maximal force). No change in serum growth hormone levels has been observed when the resistance is reduced to allow the completion of 21 repetitions. The resting level of somatotropin is not changed as a result of strength training.

Neural (Central) Factors

The central nervous system (CNS) is of paramount importance in the exertion and development of muscular strength. Muscular strength is determined not only by the quantity of involved muscle mass but also by the extent to which individual fibers in a muscle are voluntarily activated (by *intramuscular coordination*). Maximal force exertion is a skilled act in which many muscles must be appropriately activated. This coordinated activation of many muscle groups is called *intermuscular coordination*. As a result of neural adaptation, superior athletes can better coordinate the activation of fibers in single muscles and in muscle groups. In other words, they have better intramuscular and intermuscular coordination.

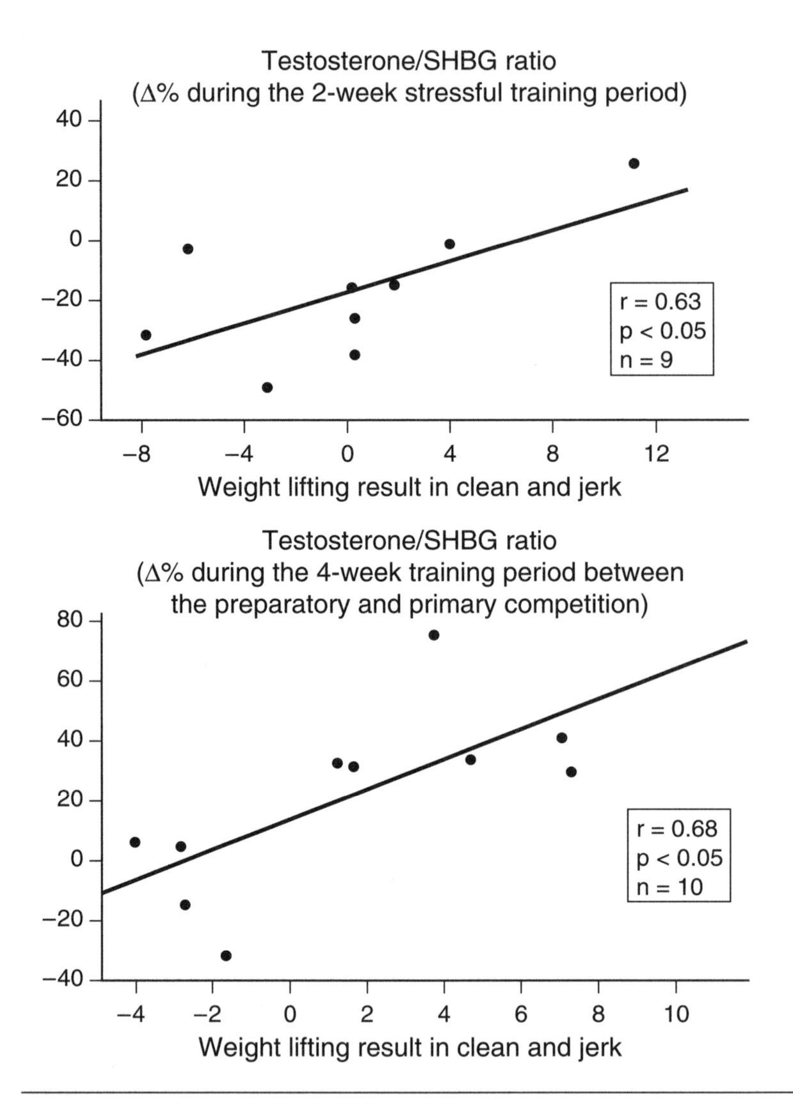

Figure 3.8 Relationships between relative changes in serum testosterone/sex hormone-binding globulin (SHBG) ratio and results of the clean and jerk lift. Testosterone is not freely soluble in plasma and must bind with plasma proteins or globulins in order to circulate in the blood. During resting conditions, more than 90% of the testosterone is bound to either SHBG or to albumin. The remaining testosterone is in a metabolically active "free" form. This study demonstrates a significant correlation between increases in strength and elevations in the free to bound testosterone ratio. The subjects were all Finnish champions or Finnish national record holders (or both) in weight lifting. Note. Adapted from "Relationships Between Training Volume, Physical Performance Capacity, and Serum Hormone Concentrations During Prolonged Training in Elite Weight Lifters" by K. Håkkinen, A. Parakinen, M. Alen, H. Kauhanen, and P. Komi, 1987, *International Journal of Sport Medicine*, **8**, pp. 61-65. Adapted by permission from the authors.

Intramuscular Coordination

The nervous system uses three options for varying muscle force production. These include

- *recruitment*, the gradation of total muscle force by the addition and subtraction of active motor units;
- *rate coding*, changing the firing rate of motor units; and
- *synchronization*, the activation of motor units in a more or less synchronized way.

All three options are based on the existence of *motor units* (MUs). MUs are the basic elements (quantums) of motor system output and consist of motoneurons, axons, motor endplates, and muscle fibers activated by one motoneuron.

MUs can be classified as *fast* or *slow* on the basis of contractile properties. *Slow MUs*, or *slow-twitch* (ST) motor units, are specialized for prolonged use at relatively slow velocities. They consist of (a) *small* low-threshold motoneurons with low discharge frequencies, (b) axons with relatively low conduction velocities, and (c) motor fibers highly adapted to lengthy aerobic activities. *Fast MUs*, or *fast-twitch* (FT) motor units, are specialized for relatively brief periods of activity characterized by large power outputs, high velocities, and high rates of force development. They consist of (a) *large* high-threshold motoneurons with high discharge frequencies, (b) axons with high conduction velocities, and (c) motor fibers adapted to explosive or anaerobic activities.

MUs are activated according to the *all-or-none law*. At any point in time, an MU is either active or inactive; there is no gradation in the level of motoneuron excitation. The gradation of force for one MU is accomplished through changes in its firing rate (rate coding).

In humans, contraction times vary from 90 to 110 ms for ST motor units and from 40 to 84 ms for FT motor units. The maximal shortening velocity of fast motor fibers is almost four times greater than the V_m of ST motor fibers. The force per unit area of fast and slow motor fibers is similar, but the FT motor units typically possess larger cross-sectionals and produce greater force per single motor unit.

All human muscles contain both ST and FT motor units. The proportion of fast and slow motor fibers in mixed muscles varies among athletes. Endurance athletes have a high percentage of ST motor units, while FT motor units are predominant among strength and power athletes.

Recruitment

During voluntary contractions, the orderly pattern of *recruitment* is controlled by the size of motoneurons (so-called *size principle*). Small motoneurons, those with the lowest firing threshold, are recruited first;

and demands for larger forces are met by the recruitment of increasingly forceful MUs. Motor units with the largest motoneurons, which have the largest and fastest twitch contractions, have the highest threshold and are recruited last. This implies, in mixed muscles containing both ST and FT motor units, that the involvement of ST motor units is forced, regardless of the magnitude of muscle tension and velocity being developed. In contrast, full FT motor unit activation is difficult to achieve. Untrained people cannot recruit all their FT motor units. Athletes engaged in strength and power training show increased MU activation.

The recruitment order of MUs is relatively fixed for a muscle involved in a specific motion, even if the movement velocity or rate of force development alters. However, the recruitment order can be changed if the multifunction muscle operates in different motions. Different sets of MUs within one muscle might have a low threshold for one motion and a high threshold for another.

The variation in recruitment order is partially responsible for the specificity of training effect in heavy resistance exercise. If the objective in training is the full development of a muscle (not high athletic performance), one must exercise this muscle in all possible ranges of motion. This situation is typical for bodybuilders and beginning athletes, but not elite athletes.

Rate Coding

The other primary mechanism for the gradation of muscle force is rate coding. The discharge frequency of motoneurons can vary over a considerable range. In general, the firing rate rises with increased force and power production.

The relative contributions of recruitment versus rate coding in grading the force of voluntary contractions are different in small and large muscles. In small muscles, most MUs are recruited at a level of force less than 50% of F_{mm}; thereafter, rate coding plays the major role in the further development of force up to F_{mm}. In large proximal muscles, such as the deltoid and biceps, the recruitment of additional MUs appears to be the main mechanism for increasing force development up to 80% of F_{mm} and even higher. In the force range between 80% and 100% of F_{mm}, force is increased almost exclusively by intensification of the MU firing rate.

Synchronization

Normally, MUs work asynchronously to produce a smooth, accurate movement. However, there is some evidence that, in elite power and strength athletes, MUs are activated synchronously during maximal voluntary efforts.

In conclusion, maximal muscular force is achieved when

1. a maximal number of both ST and FT motor units are recruited;
2. rate coding is optimal to produce a fused tetanus in each motor fiber; and
3. the MUs work synchronously over the short period of maximal voluntary effort.

Psychological factors are also of primary importance. Under extreme circumstances (i.e., "life-or-death" situations), people can develop extraordinary strength values. When untrained subjects (but not superior athletes) receive hypnotic suggestions of increased strength, they exhibit strength increases, whereas both athletes and untrained people show strength decrements after receiving the hypnotic suggestion of decreased strength. Such strength enhancement is interpreted to mean that the CNS in extraordinary situations either increases the flow of excitatory stimuli, decreases the inhibitory influence to the motoneurons, or both.

It may be that the activity of motor neurons in the spinal cord is normally inhibited by the CNS and that it is not possible to activate all MUs within a specific muscle group. Under the influence of strength training and in exceptional circumstances (important sport competitions included), a reduction in neural inhibition occurs with a concomitant expansion of the recruitable motoneuron pool and an increase in strength.

Intermuscular Coordination

Every exercise, even the simplest one, is a skilled act requiring the complex coordination of numerous muscle groups. The entire movement pattern, rather than the strength of single muscles or the movement of single joints, must be the primary training objective. Thus, an athlete should use "local" strength exercises, in which the movement is performed in only one joint, and only as a supplement to the main training program.

Here are some examples of the primary importance of the entire coordination pattern (rather than the force of single muscles) for muscular strength.

- *Electrostimulation training*. It is possible to induce hypertrophy and increase the maximal force of a single muscle, for instance the rectus femoris, or even a muscle group (e.g., knee extensors), through electrostimulation (EMS). However, if only EMS is used, it takes a great deal of time and effort to transmute this increased potential into a measurable strength gain in a multijoint movement such as a leg extension. Some athletes who try EMS decide that it is not worth the effort (see also chapter 6 on EMS). Strength gains attained through conventional voluntary training rely on changes in the nervous system that do not occur when muscles are stimulated electrically.

- The best weight lifters are the strongest people in the world, but they cannot perform slow gymnastic exercises, which require "only" strength (e.g., the "cross" on the rings). On the other hand, elite gymnasts do not exercise with free weights to increase the force of the shoulder girdle muscles. They do this with gymnastic exercises using body weight as resistance (heavy ankle cuffs or waist belts are added from time to time).
- *Bilateral deficits.* If an athlete simultaneously exerts maximal force with two extremities, the force for each extremity is lower than it is in unilateral force development. Training with bilateral contractions reduces the bilateral deficit. Athletes in sports such as rowing or weight lifting that require the simultaneous bilateral contraction of the same muscle groups should use similar exercises to eliminate bilateral deficits. (However, the elite super heavy lifters employ exercises such as stepping up on a bench with barbells 180 kg and heavier; they do this to avoid the extremely high loading that occurs during squatting exercises, in which the barbell weight can exceed 350 kg.)

In the case of the *bottleneck effect,* when low strength in one joint of a kinematic chain limits performance (e.g., knee extensor strength is the limiting factor in squatting), the coach should first try to change the exercise to redistribute the load among different muscle groups. Only after that is an isolated knee extension against a resistance advisable.

The important limitation of many strength training machines is that they are designed to *train muscles,* not movement. Because of this, they are not the most important training tool for athletes.

Taxonomy of Strength

Let us review some facts from chapters 2 and 3:

1. Magnitudes of the maximal force F_m in slow movements do not differ greatly from those in isometric actions.
2. The greatest muscular forces are developed in eccentric actions; such forces are sometimes twice those developed in isometric conditions.
3. In concentric actions, the force F_m is reduced when the time to peak force T_m decreases or the velocity increases.
4. There are no substantial correlations between maximum maximorum force (F_{mm}) and the force F_m in movements with minimal external resistance (note, body weight is not minimal resistance). The correlation is greater when the resistance is increased.

5. The rate of force development (especially the S-gradient) does not correlate with the maximal force F_{mm}.
6. The force in exercises with reversible muscle action does not change after heavy resistance training, regardless of the F_{mm} increase (this is true at least for experienced athletes).

In summary, the following general scheme can be proposed as a *taxonomy of muscular strength*:

Type of Strength	*Manifestation*
Static strength (or, simply, strength)	Isometric and slow concentric actions
Dynamic strength	Fast concentric movements
Yielding strength	Eccentric actions

Additionally, the explosive strength (or *rate of force development*) and the force exerted in stretch-shortening (*reversible*) muscle actions are considered independent components of motor function.

The summary classification scheme is certainly not completely satisfactory from a scientific point of view in that it uses different bases for categorization (direction of movement, velocity, time). Furthermore, a smooth transition exists rather than a sharp demarcation between different types of strength. Despite these valid criticisms, this classification system has served as a useful tool in practical work for many years. Unfortunately, a better system does not exist at this time.

Summary

To understand what determines the differences across athletes, we scrutinize two factors: peripheral (that is, capabilities of individual muscles) and central (the coordination of muscle activity by the CNS). Among *peripheral factors*, muscle dimensions seem to be the most important: Muscles with a large physiological cross-sectional area produce higher forces. The size of a muscle increases when (a) a properly planned strength training program is executed and (b) the required amount and selection of amino acids are provided via nutrition. The enlargement of the cross-sectional area of individual fibers (fiber hypertrophy) rather than an increase in the number of fibers (hyperplasia) is responsible for muscle size growth. Heavy resistance exercise activates the breakdown of muscle proteins, creating conditions for the enhanced synthesis of contractile proteins during rest periods. The mass of proteins catabolized during exercise exceeds the mass of newly synthesized protein. The crucial factor for increasing the protein breakdown is a shortage in the muscle cell of energy available for protein buildup during heavy resistance exercise.

Since muscle mass constitutes a substantial part of the human body, athletes with larger body weight demonstrate greater strength than equally trained athletes of smaller body dimensions. The strength per kilogram of body weight is called *relative strength*; muscular strength, when not related to body weight, is termed *absolute strength*. Among equally trained athletes of various weight classes, absolute strength increases and relative strength decreases with a gain in body weight. Body weight loss, if properly managed, is helpful toward increasing relative force. However, athletes must be warned against the malpractice of rapid weight reduction.

Neural (central) factors include intramuscular and intermuscular coordination. On the level of intramuscular coordination, three main options are used by the CNS for varying muscle force production: recruitment of MUs, rate coding, and synchronization of MUs. These can be observed in well trained athletes during maximal efforts. The orderly recruitment of MUs is controlled by the size of motoneurons (Hennemann's size principle): Small motoneurons are recruited first and requirements for higher forces are met by the activation of the large motoneurons that innervate fast MUs. It seems that the involvement of slow twitch MUs is forced, regardless of the magnitude of muscle force and velocity being developed. The firing rate of the MUs rises with increased force production (rate coding). The maximal force is achieved when (a) a maximal number of MUs is recruited, (b) rate coding is optimal and (c) MUs are activated synchronously over the short period of maximal effort.

The primary importance of intermuscular coordination for generating maximal muscular force is substantiated by many investigations. Thus, the entire movement pattern rather than the strength of individual muscles or single joint movements should be the primary training objective. Explosive strength (or rate of force development) and the force exerted in stretch-shortening (reversible) muscle actions are independent components of motor function.

PART II

METHODS OF STRENGTH CONDITIONING

Part 2 summarizes the requisite knowledge for coaching successfully, concentrating on information derived both from scientific evidence and the documented practical experience of elite athletes. Chapter 4, which covers intensity and methods of strength training, begins with the description of measurement techniques. It also reviews current scientific material about exercising with different resistance, analyzing metabolic reactions, intramuscular coordination, and biomechanical variables. The chapter then scrutinizes the training intensity of elite athletes and presents data from the training logs of dozens of the best athletes in the world, including Olympic champions and world-record holders from Eastern Europe. The best athletes in these countries tend to train together and thus lend themselves to easier monitoring, and you will see reflected in this book some 35 years' worth of training logs of many sport stars. Chapter 4 also outlines three main methods of strength training and discusses in detail the parallels between practical training and scientific lore.

Chapter 5 turns to the aspects of timing during training, including short-term and medium-short-term timing. It covers the main problems of short-term planning; how to use strength exercises in workouts and training days, as well as in micro- and mesocycles; and the four main aspects of periodization: delayed transformation, delayed transmutation, training residuals, and the superposition of training effects.

Chapter 6 pertains to the issue that coaches face first and foremost when they devise strength training programs: exercise selection. The chapter examines various strength exercises; it also classifies exercises and presents a rationale for exercise selection. For experienced athletes, decisions are fairly complex, and among the exercise features they must consider are the following: working muscles, type of resistance, time and rate of force development, movement velocity, movement direction, and the force-posture relationship. Chapter 6 also describes the peak-contraction principle, accommodating resistance, and accentuation—the three basic techniques used in modern strength training to handle the force-posture paradigm.

A later section of chapter 6 concentrates on strength exercises that are regarded by many as supplementary, including isometric exercises, self-resistance exercises, and yielding exercises. We note that exercises with reversible muscle action, such as dropping jumps, are becoming more popular. Meanwhile, the sport exercises that call for added resistance, which are often referred to as "speed-resisted," can now hardly be called auxiliary. In fact, some experts see the shift in popularity of this group of exercises as the most visible trend in training during the 1980s. Chapter 6 explains how to choose and use all these training techniques. It then reviews electrostimulation—a training technique that is sometimes labeled exotic—and ends by offering some practical advice on how to breathe while exercising.

Chapter 7 describes measures that may prevent injuries during strength training, especially to the lumbar region, explaining the underlying theory while presenting practicable techniques. Several applied aspects are discussed, including muscle strengthening, sport technique requirements, use of protective implements, posture correction and flexibility development, and rehabilitation measures.

Chapter 8 explores goal-specific strength training. Both athletes and lay people exercise for strength not only to improve strength performance but also for many other reasons (goals may be as diverse as power performance, muscle mass gain, endurance performance, or injury prevention). The chapter also summarizes specific features of strength training.

CHAPTER 4

Training Intensity: Methods of Strength Training

In this chapter we turn to the topic of training intensity and focus on four major issues. First we consider several methods of measuring training intensity. Then we look at the physiological characteristics of exercises with varying intensities, particularly the influence of different strength exercises on metabolism and intra- and intermuscular coordination. The third issue is the training intensities of elite, world-class athletes—information that suggests which training patterns are the most efficient. The final section outlines the theory underlying a described training pattern and presents the primary methods of strength training.

Measurement Techniques

Training intensity can be estimated in four different ways: These are by the

- *magnitude of resistance* (e.g., weight lifted) expressed as a percentage of the best achievement (F_m or F_{mm}) in a relevant movement (if the weight lifted is expressed in kilograms, it is difficult to compare the training loads of athletes who vary in mastership and weight class);
- *number of repetitions* (lifts) per set (a set is a group of repetitions performed consecutively);
- *number (or percentage) of repetitions* with maximal resistance (weight); and
- *workout density,* i.e., the number of sets per hour in a workout.

You can read in this chapter about the first three methods; and in chapter 5 about workout density.

To characterize the *magnitude of resistance* (load), we use the percentage of weight lifted relative to the athlete's best performance. Depending on how the "best achievement" is determined, two main variants of such a measure are utilized. One option is to use the athletic performance attained during an official sport competition (*competition* F_{mm} = CF_{mm}). The second option is to use a maximum training weight (TF_{mm}) for comparison.

By definition, *maximum training weight* is the heaviest weight (one repetition maximum) an athlete can lift without substantial *emotional* stress. In practice, experienced athletes determine TF_{mm} by registering heart rate. If the heart rate increases before the lift, this is a sign of emotional anxiety. The weight exceeds TF_{mm} in this case. (Note, however, that heart rate elevation before lifting the maximal competition load CF_{mm} varies substantially among athletes. During important competitions, the range is between 120 and 180 beats per minute. To determine TF_{mm}, athletes must know their individual reactions.) The difference between the TF_{mm} and the CF_{mm} is approximately 12.5 ± 2.5% for superior weight lifters. The difference is greater for athletes in heavy weight classes. For athletes who lift 200 kg during competition, a 180-kg weight is typically above their TF_{mm}.

For an athlete, the difference between CF_{mm} and TF_{mm} is great. After an important competition, weight lifters are extremely tired even though they may have performed only 6 lifts, in comparison to nearly 100 lifts during a regular training session. Such athletes have a feeling of "emptiness" and cannot lift large weights. Thus they need about 1 week of rest and cannot compete in an important competition until 1 month of rest and training has expired (compared with the situation in other sports in which contests are held two to three times a week). The reason is not the physical load itself but the great emotional stress an athlete experiences while lifting CF_{mm}. TF_{mm} can be lifted at each training session.

It is more practical to use CF_{mm} than to use TF_{mm} for the calculation of training intensity. Since the 1960s, the average training intensity for elite soviet athletes has been 75 ± 2%. Superior weight lifters do not necessarily exercise with such intensity. Athletes from other countries often use higher or lower training weights. For instance, in 1987, Finnish weightlifting champions exercised with an average intensity of 80 ± 2.5%. Soviet or Russian athletes, however, as well as Bulgarians, have won nearly all the gold medals at the world and Olympic championships over the past 25 years.

In a sport such as weight lifting, the training intensity is characterized also by an *intensity coefficient*. This ratio is calculated:

$$\frac{\text{Average weight lifted, kg} \cdot 100}{\text{Athletic performance (snatch plus clean and jerk), kg}}$$

On average, the intensity coefficient for superior soviet athletes has been 38 ± 2%.

It is a good idea to use as a CF_{mm} value the average of two performances attained during official contests immediately before and after the period of training you are studying. For instance, if the performance was 100 kg during a competition in December and it was 110 kg in May, the average CF_{mm} for the period January through April was 105 kg. There are many misconceptions in sport science literature regarding the training loads used in heavy resistance training. One reason is that the difference between CF_{mm} and TF_{mm} is not always completely described. It is important to pay attention to this difference.

The number of repetitions per set is the most popular measure of exercise intensity in situations in which maximal force F_{mm} is difficult or even impossible to evaluate, for instance in sit-ups.

The magnitude of resistance (weight, load) can be characterized by the ultimate number of repetitions possible in one set (to failure). The maximal load that can be lifted a given number of repetitions before fatigue is called *repetition maximum* (RM). For instance, 3 RM is the weight that can be lifted in one set only three times. Determining RM entails the use of trial and error to find the greatest amount of weight a trainee can lift a designated number of times. RM is a very convenient measure of training intensity in heavy resistance training. However, there is no fixed relationship between the magnitude of the weight lifted (expressed as a percentage of the F_{mm} in relevant movement) and the number of repetitions to failure, RM. This relationship varies with different athletes and motions (Figure 4.1). As the figure shows, 10 RM corresponds to approximately 75% F_m. This is valid for athletes in sports where strength and explosive strength predominate (such as weight lifting, sprinting, jumping, and throwing). However, note that a given percentage of 1 RM will not always correspond to the same number of repetitions to failure in the performance of different lifts.

During training, superior athletes use varying numbers of repetitions in different lifts. In the snatch and the clean and jerk the typical number of repetitions ranges from one to three, and the most common number is two (almost 60% of all sets are performed with two repetitions). In barbell squats, the range is from two to seven lifts per set (more than 93% of all sets are performed in this range; see Figure 4.2). You will find further examples and an explanation of these findings later on in this chapter. As a rule of thumb, fewer than 10 to 12 RM should be used for muscular strength development; the exceptions to this are rare (e.g., sit-ups).

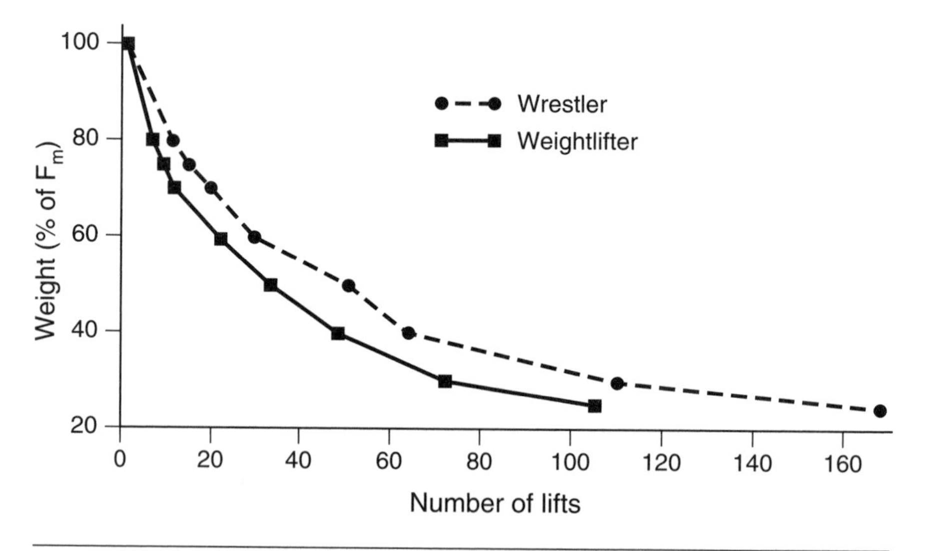

Figure 4.1 Dependence of the maximal number of repetitions to failure (RM, abscissa) to weight lifted (% F_{mm}, ordinate). Bench press; the results for two qualified athletes, a weight lifter and a wrestler, are shown. The pace of lifts was 1 lift in 2.5 s. Both athletes were highly motivated. Note. The data are from "Relationships Between the Motor Abilities" by V.M. Zatsiorsky, N.G. Kulik, and Yu. I. Smirnov, 1968, *Theory and Practice of Physical Culture*, **31**(12), pp. 35-48; 1969, **32**(1), pp. 2-8; 1969, **32**(2), pp. 28-33. Reprinted by permission from the journal.

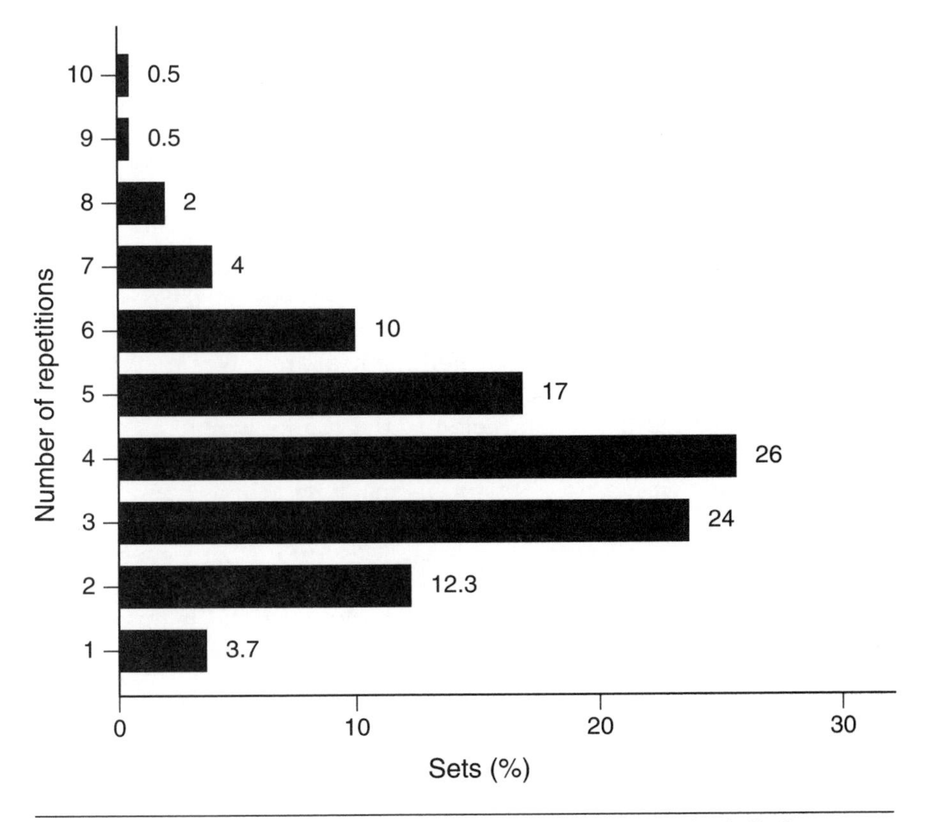

Figure 4.2 Number of repetitions per set in squatting with a barbell. One year of observations in the training of eight world and Olympic champions in the clean and jerk lift. Note. The data are from *Preparation of National Olympic Team in Weight Lifting to the 1988 Olympic Games in Seoul,* 1969, technical report #1988-67 (p. 79), Moscow: All-Union Research Institute of Physical Culture.

The number of repetitions with maximal resistance is used as an additional, but very useful, measure of the intensity of strength training. By agreement, all lifts with a barbell above 90% of CF_{mm} are included in this category. These loads are above TF_{mm} for almost all athletes.

Determination of Training Intensity

A conditioning coach wants to prescribe a training intensity in barbell squats for two athletes, A and B. Athlete A is a competitive weight lifter from a light weight class; athlete B is a football player. Recently during a modeled competition, athlete A managed to lift a 150-kg barbell (his CF_{mm}). To prepare for the competition, athlete A excluded barbell squats from his training program for 10 days before the contest and had a complete 2-day rest. He considered the competition very important and psychologically prepared himself to set his best personal achievement in squatting. During the competition, athlete A performed squats in a fresh condition, immediately after a warm-up. Because of high emotional stress, his heart rate before the lifts was approximately 180 beats/min.

For this athlete, the maximal training weight must be around 135 kg—his TF_{mm}. To define this weight more precisely, the coach monitored the athlete's heart rate during rest intervals and found that before he lifted a 135-kg barbell his heart rate was not elevated. Therefore this weight did not elicit high emotional stress. The coach recommended that the athlete use the 135-kg weight as maximal load in the majority of training sessions during the next training cycle. This was exactly 90% of his maximal achievement attained during the competition.

Without experiencing emotional stress and using a special competition-like warm-up, athlete A was able to lift a 135-kg barbell one to two times in one set. Since the advice was to perform three to four squats in a set, athlete A exercised mainly with the 125–130-kg barbell. Periodically, he also used higher loads, including some greater than 135 kg. These lifts were counted and their numbers used as an additional measure of training intensity.

Athlete B also squatted with a 150-kg barbell. But unlike athlete A, he did this during a regular session within his usual training routine. Additional rest before the test was not provided and no special measures were taken. For this athlete, the 150-kg achievement can be regarded as a maximum training weight (TF_{mm}). He can exercise with such a load regularly.

Exercising With Different Resistance

Different levels of resistance have different physiological effects. Varied resistance levels cause different *metabolic reactions* involving the breakdown and synthesis of proteins. The resistance level or intensity of exercise also influences *intramuscular* and *intermuscular* coordination.

Metabolic Reactions

According to the energetic hypothesis of muscle cell hypertrophy described earlier (see chapter 3, muscle dimensions), the crucial factor determining the balance between protein catabolism and anabolism is the amount of energy available for protein synthesis during exercise. If the resistance is relatively small, the energy available in the muscle cell is conveyed for muscle action and, at the same time, for anabolism of muscle proteins. Thus, the energy supply satisfies both requirements. During heavy weight lifting, a larger amount of energy is provided to the contractile muscle elements and spent on muscular work. Energy transfer for the synthesis of proteins decreases, while the rate of protein breakdown (the amount of degraded protein per lift) increases. The rate of protein degradation is a function of the weight lifted: The heavier the weight, the higher the rate of protein degradation.

The total amount of degraded protein, however, is a function of both the rate of protein catabolism and the mechanical work performed (or the total weight lifted). Mechanical work is greater when resistance is moderate and several consecutive lifts are performed in one set. For instance, if an athlete presses a 100-kg barbell 1 time (this athlete's RM), the total weight lifted is also 100 kg. However, the same athlete should be able to lift a 75-kg barbell (to failure) about 10 times; here the total weight lifted equals 750 kg.

The mass of proteins catabolized during heavy resistive exercise can be presented as a product of the rate of protein breakdown and the number of lifts. If the resistance is very large (e.g., 1 RM), the rate of protein breakdown is high but the number of repetitions is small. At the other extreme, if the resistance is small (50 RM), the number of lifts and amount of mechanical work are great, but the rate of protein degradation is low. So the total amount of the degraded protein is small in both cases, but for different reasons (Table 4.1).

An additional feature of such training, an important one from a practical standpoint, is the very high training volume or total amount of weight lifted during a workout. This amount is up to five or six times greater than the amount lifted during a conventional training routine. Athletes

Table 4.1 The Amount of Degraded Protein During Strength Training With Different Levels of Resistance

Resistance, RM	Rate of protein degradation	Mechanical work (number of repetitions)	Total amount of of degraded protein
1	High	Small	Small
5–10	Average	Average	Large
>25	Low	Large	Small

who train over a certain period of time in this manner (to gain body weight and induce muscle cell hypertrophy in order to compete in a heavier weight class) amass a training volume in one workout of over 20 to 30 tons and, in some cases, above 50 tons a day. Such volume hinders an athlete's capacity to perform other exercises during this period of training.

■ *Exercising With Various Weights: Mechanical Work and Metabolic Response*

An athlete whose best achievement in barbell squatting is 150 kg performs squats with 150-, 120-, and 90-kg barbells. His body weight is 77.5 kg and the weight of body parts above the knee joints is 70 kg (only this part of the body is lifted during squatting; the feet and shanks are almost motionless). Thus, the weights lifted (the barbell plus the body) are 220, 290, and 160 kg. The distance that the center of gravity is raised (the difference between the lowest and the highest position of the center of gravity) is 1 m. The athlete lifts the 150-kg barbell 1 time, the 120-kg barbell 10 times, and the 90-kg barbell 25 times. The mechanical work produced equals 220 kgm for the heaviest barbell (220 kg multiplied by 1 time and 1 m), 1,900 kgm for the 120-kg barbell, and 4,000 kgm for the lightest one (160 kg · 25 times · 1 m). Exercising with a light barbell, the athlete produces mechanical work 18 times greater than with the heaviest.

The metabolic energy expenditures are many times larger during exercise with the light barbell. However, protein degradation is maximized when squats are performed with the 120-kg ("average") barbell. During squatting with the 150-kg barbell, the intensity of protein catabolism (the amount of degraded proteins per repetition) is very high. This barbell, however, is lifted one time only. When the athlete executes the squats with the light (90 kg) barbell, the intensity of protein degradation is low. So the amount of degraded protein is low in spite of the huge value of mechanical work produced. Thus, the 120-kg load provides this particular athlete with the best combination of training intensity and volume (total load lifted).

Intramuscular Coordination

Lifting maximal weight has a number of effects on motor units (MUs): A maximum number of MUs are activated, the fastest MUs are recruited, the discharge frequency of motoneurons is at its highest, and the activity of MUs is synchronous.

MUs do exist, however, that many athletes cannot recruit or raise to the optimal firing rate in spite of sincere efforts to develop maximal force. It is well known that high-threshold (fast) units possess a higher maximal discharge frequency. However, investigators have shown that, in untrained people during maximal voluntary contractions, many high-threshold MUs exhibit a lower firing frequency than low-threshold MUs. This is so because the fast MUs are not fully activated even though the individual is attempting to attain maximal forces.

The "hidden potential" of a human muscle to develop higher forces can also be demonstrated by electrostimulation. In experiments involving maximum voluntary contraction, the muscle is stimulated with electrical current (for detail on muscle electrostimulation see chapter 6). The stimulus induces an increase in force production. The ratio

$$\frac{(\text{Force during electrostimulation} - \text{Maximal voluntary force}) \cdot 100}{\text{Maximal voluntary force}}$$

is called the *muscle strength deficit* (MSD). The MSD typically falls in the range of 5% to 35%. The MSD is smaller for elite athletes; it is smaller also when a person is anxious or when only small muscles are activated. The very existence of the MSD indicates that human muscles typically have "hidden reserves" for maximal force production that are not used during voluntary efforts.

One objective of heavy resistance training is to "teach" an athlete to recruit all the necessary MUs at a firing rate that is optimal for producing a fused tetanus in each motor fiber. When submaximal weights are lifted, an intermediate number of MUs are activated; the fastest MUs are not recruited; the discharge frequency of the motoneurons is submaximal; and MU activity is asynchronous. It is easy to see differences in intramuscular coordination between exercises with maximal versus submaximal weight lifting. Accordingly, exercises with moderate resistance are not an effective means of training for strength development, particularly when improved intramuscular coordination is desired.

Many people believe that, in the preparation of elite weight lifters, optimal intramuscular coordination is realized when weights equal to or above TF_{mm} are used in workouts. It is not mandatory from this standpoint to lift CF_{mm} during training sessions. Differences in the best performances attained during training sessions (i.e., TF_{mm}) and during important competition (i.e., CF_{mm}) are explained by psychological factors such as the level of arousal and by increased rest before a contest (recall the two-factor theory of training in chapter 1). Differences in coordination (intra- and intermuscular), however, do not affect performance. Weights above TF_{mm} are used only sporadically in training (for approximately 3.5–7.0% of all the lifts).

■ *What Happens When a Nonmaximal Load Is Lifted?*

A person curls a 30-kg dumbbell, causing the following to occur: (a) the maximal number of MUs are recruited; (b) the fast MUs, which are also the strongest, are activated; (c) the discharge frequency of motoneurons is optimal; and (d) motoneuron activity is (maybe) synchronous.

However, when a 15-kg dumbbell is lifted, (a) only a portion of the total MUs are recruited, (b) the fastest (and strongest) MUs are not activated, (c) the frequency of neural stimulation is not optimal, and (d) MU activity is (surely) asynchronous.

Intramuscular coordination in the two activities is substantially different. Thus, lifting a 15-kg load cannot improve the intramuscular coordination required to overcome a 30-kg resistance.

Biomechanical Variables and Intermuscular Coordination

When an athlete lifts maximal weights, movement velocity reaches its ultimate value and then remains nearly constant. Acceleration of the barbell varies near the zero level and the force is more or less equal to the weight of the object lifted (Figure 4.3a).

In the lifting of moderate weights, there can be two variations. In the first example (Figure 4.3b), efforts are maximally applied. Acceleration increases in the initial phase of the lift, then falls to zero and becomes negative in the second phase of the motion. At the beginning the force applied to the barbell is greater than the weight lifted and then decreases. The second part of the motion is partially fulfilled via the weight's kinetic energy. In this type of lifting, muscular coordination differs from that utilized in the lifting of maximal or near-maximal weights. That is, muscular efforts are concentrated ("accentuated") only in the first half of the movement.

In the second instance, kinematic variables of the movement (velocity, acceleration) are similar to those observed when a person does a maximal lift. Acceleration, and the corresponding external force applied to the barbell, are almost constant. However, this motion pattern—the intentionally slow lift—involves the coactivation of antagonistic muscle groups. Such intermuscular coordination hampers the manifestation of maximum strength values.

Differences in underlying physiological mechanisms, experienced when exercising with various loads, explain why muscular strength increases only when exercises requiring high forces are used in training.

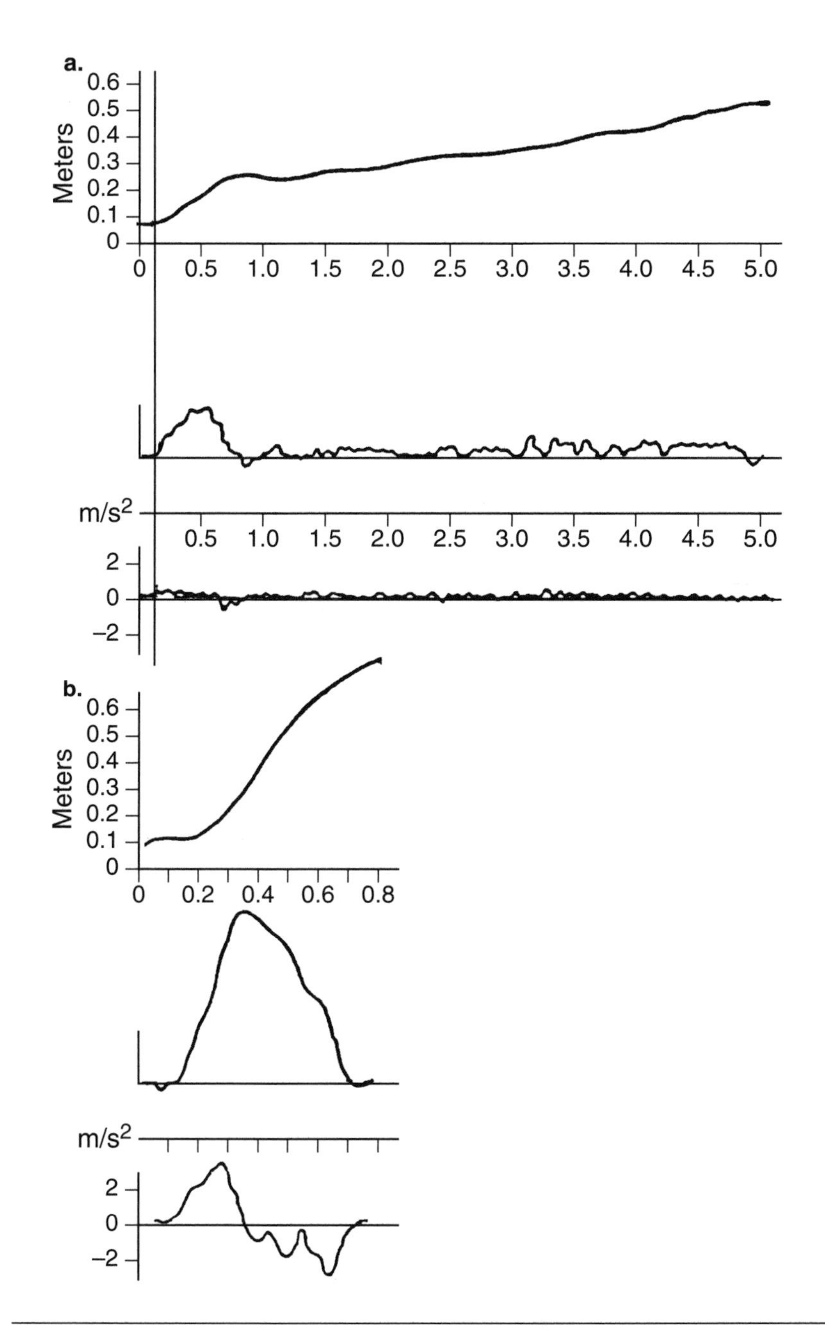

Figure 4.3 Distance (m), velocity (m/s), and acceleration (m/s^2) during a military barbell press. (a) maximal weight; (b) small weight. Note. From *Motor Abilities of Athletes* (p. 30) by V.M. Zatsiorsky, 1966, Moscow: FiS. Reprinted by permission from Fizkultura i Sport.

In principle, workloads must be above those normally encountered. The resistance against which the muscle groups work must continually be increased as strength gains are made (this is called *the principle of progressive-resistance exercises*).

In untrained individuals, the measured strength levels fall when resistance is below 20% of their F_{mm}. In athletes accustomed to great muscular efforts, this drop in strength can begin even with loads that are relatively heavy, but below their usual level. For instance, if qualified weight lifters train with weights of 60% to 85% TF_{mm} and do not lift these loads in one set to failure (to fatigue), the strength level is kept constant over the 1st month of such training and drops 5% to 7% during the 2nd month. Athletes in seasonal sports, such as rowing, lose strength levels previously attained in the preparation period if they do not use high-resistance training during a competition period, regardless of intense specific workouts.

Qualified athletes retain only muscle size, not muscular strength, when they use exclusively moderate (nonmaximal) resistances and moderate (nonmaximal) repetitions over a period of several months.

Training Intensity of Elite Athletes

The practical training experience of elite athletes is a useful source of information in sport science. This experience, although it does not provide sound scientific proof of the optimality of the training routines employed, reflects the most efficient training pattern known at the present time. In the future, gains in knowledge will certainly influence training protocols. Currently, however, we do not know precisely what the best approaches are.

The *distribution of training weights* in the conditioning of elite weight lifters is shown in Figure 4.4. Notice that elite athletes use a broad spectrum of loads. They use loads below 60% of CF_{mm} mainly for warming up and restitution (these loads account for 8% of all the lifts). The highest proportion of weights lifted (35%) consists of those 70% to 80% of the CF_{mm}. In agreement with these data and as observed over many years, the average weight lifted by superior athletes is equal to 75.0 ± 2.0% of CF_{mm}. Loads above 90% of CF_{mm} account for only 7% of all lifts.

The *number of repetitions per set* varies by exercise. In both the snatch and the clean and jerk lifts (Figure 4.5), the majority of all sets are performed with one to three repetitions. In the snatch, only 1.8% of the sets are done with three or four repetitions; in the clean and jerk, the percentage of sets with four to six lifts is no more than 5.4%. The majority of sets, roughly 55% to 60%, consist of two repetitions.

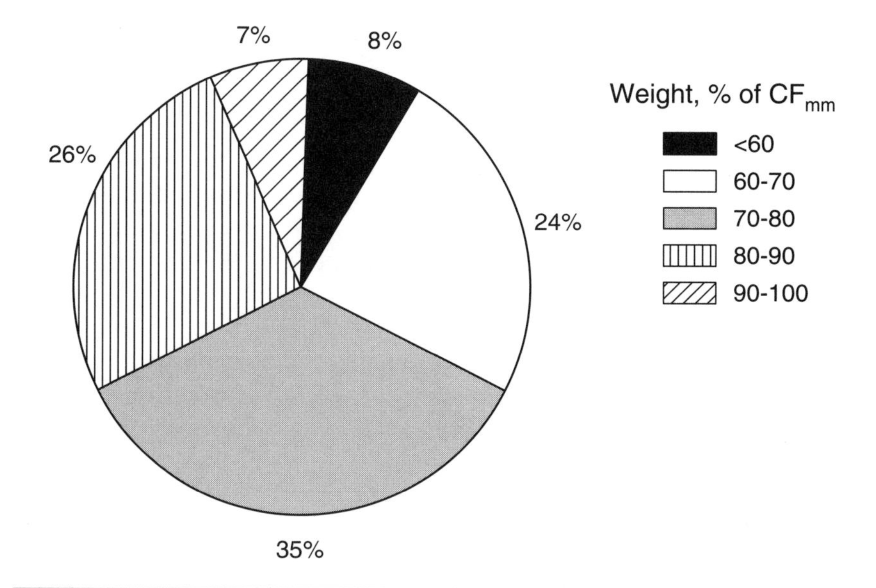

Figure 4.4 The distribution of weights lifted by members of the U.S.S.R. Olympic team during preparation for the 1988 Olympic Games. The exercises were divided into two groups: snatch related and clean related. The weight lifted is expressed as a percentage of CF_{mm} in main sport exercises (either in snatch or clean and jerk). Squatting with the barbell is not included in this analysis. One year of direct observations. Note. Adapted from *Training Load in Strength Training of Elite Athletes* by V.M. Zatsiorsky, 1991, Oct. 26-31, paper presented at the Second IOC World Congress on Sport Sciences, Barcelona.

In auxiliary strength exercises such as squatting with a barbell, where motor coordination only partially resembles the coordination in the snatch and clean and jerk, the number of repetitions in one set increases (recall Figure 4.2). In barbell squats, for instance, the number of lifts varies from 1 to 10, with the average range being 3 to 6.

Generally speaking, as the intermuscular coordination in an exercise becomes more simple and as the technique of the exercise deviates from the technique of the main sport event (in this example, from the technique of both the snatch and clean and jerk), the number of repetitions increases. In the clean and jerk, it is one to three (57.4% of sets contain two lifts only); the typical number of repetitions in squatting is three to five, and in the inverse curl the average number of lifts is around five to seven per set (Figure 4.6).

The *number of repetitions with maximal resistance* (near CF_{mm}) is relatively low. During the 1984–1988 Olympic training cycle, elite soviet athletes lifted a barbell of such weight in main sport exercises (snatch, clean and

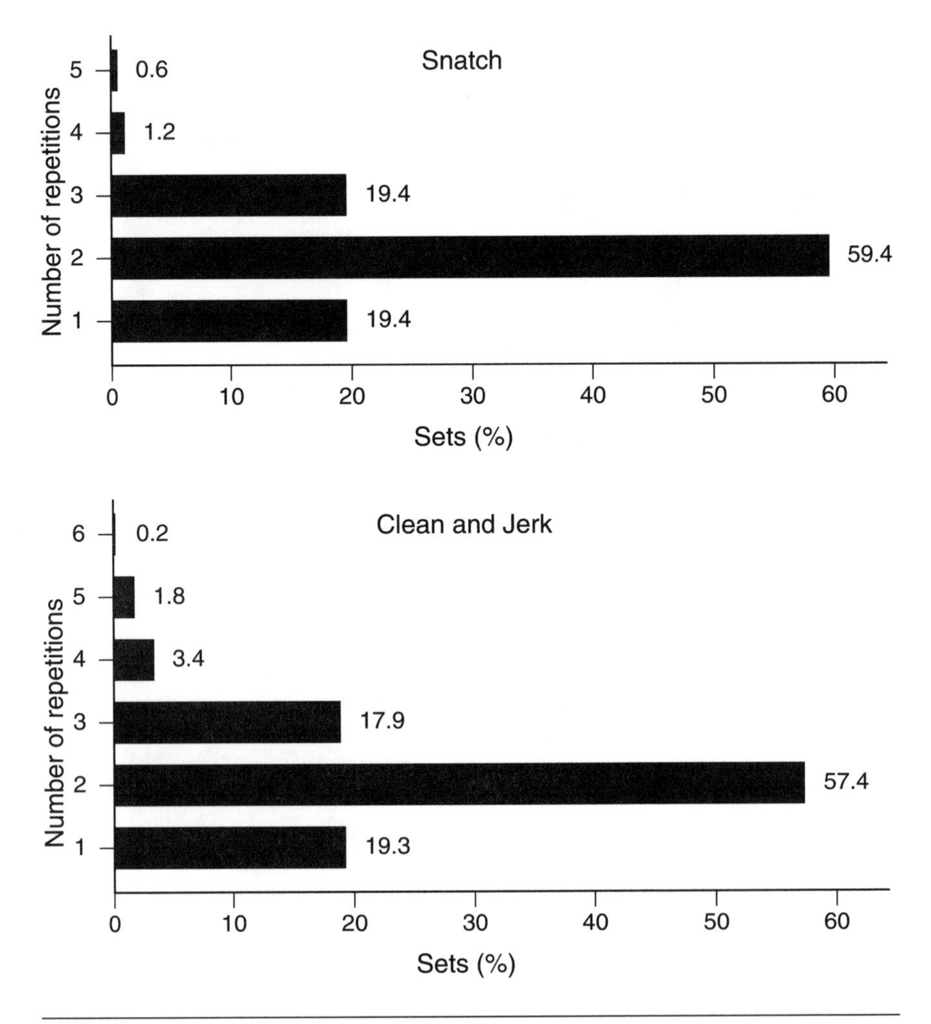

Figure 4.5 The percentage of sets with various numbers of lifts in the training of superior athletes. Note. Adapted from *Training Load in Strength Training of Elite Athletes* by V.M. Zatsiorsky, 1991, Oct. 26-31, paper presented at the Second IOC World Congress on Sport Sciences, Barcelona.

jerk) 300 to 600 times a year. This amount comprised 1.5 to 3.0% of all performed lifts. The weights were distributed as follows:

Weight of barbell, % of CF_{mm}	***Number of lifts, %***
90–92.5	65
92.6–97.5	20
97.6–100	15

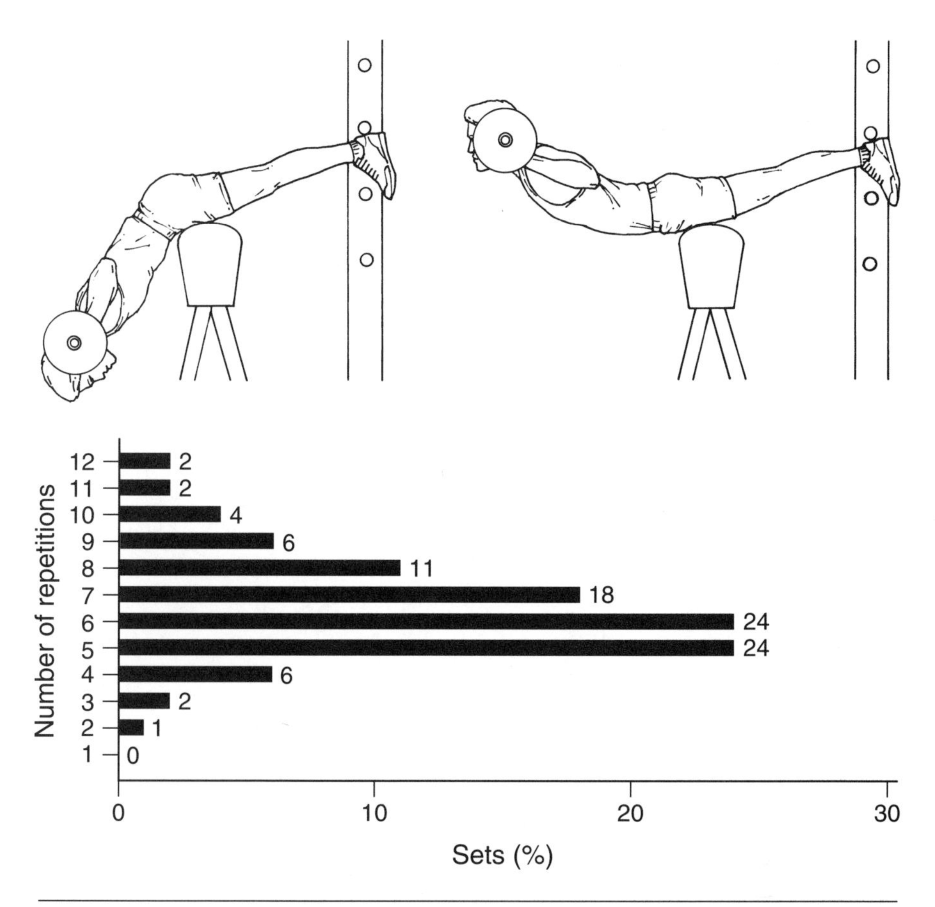

Figure 4.6 Inverse curl (top) and the percentages of sets with different numbers of lifts in this exercise. Results of direct observations in the training of V. Alexeev, many-time world and Olympic champion and world record holder in weight lifting (super heavy weight category) in a total of 130 sets. Note. The data are from *Preparation of National Olympic Team in Weight Lifting to the 1980 Olympic Games, in Moscow*, 1981, technical report #1981-34, Moscow: All-Union Research Institute of Physical Culture.

In the 1-month period before important competitions, weights above 90% of CF_{mm} are lifted in the snatch or clean and jerk, or a combination of the two, 40 to 60 times.

During the 1980s, the soviet and Bulgarian weight-lifting teams won almost all gold medals at the world and Olympic competitions. It has often been reported that Bulgarian weight lifters lift barbells of maximal weight more than 4,000 times a year. The training intensity of Bulgarian athletes is actually higher than it is for soviet athletes. However, the real source of such a huge discrepancy (600 vs. 4,000 lifts a year) is not the training itself, but the method of determining maximal weight. In their plans and logs soviet athletes use CF_{mm}, while Bulgarians stick to the TF_{mm} designation (1 RM in a given training session).

These repetition levels should not be mechanically copied. Rather, coaches and athletes need to understand the ideas underlying such training (*the training conception*) and, if they accept this conception, thoughtfully implement it. The training conception described in the following paragraphs includes understanding specific features of different training methods, proper exercise selection, and timing of training.

Methods of Strength Training

It is useful to classify strength training according to methods of attaining maximal muscular tension. In the literature, methods of strength training are sometimes classified according to the exercises used (isometric, isotonic, eccentric, etc.). We prefer to use this classification as a taxonomy of strength exercises rather than training methods. There are three ways to achieve maximal muscular tension:

1. Lifting a maximum load (exercising against maximal resistance)—that is, the *maximal effort method*
2. Lifting a nonmaximal load to failure; during the final repetitions the muscles develop the maximum force possible in a fatigued state—the *repeated effort method*
3. Lifting (throwing) a nonmaximal load with the highest attainable speed—the *dynamic effort method*

In addition, the lifting of nonmaximal loads an intermediate number of times (not to failure) is used as a supplementary training method (the *submaximal effort method*).

Maximal Effort Method

The method of maximal effort is considered superior for improving both intramuscular and intermuscular coordination; the muscles and central nervous system (CNS) adapt only to the load placed upon them. This method should be used to bring forth the greatest strength increments. CNS inhibition, if it exists, is reduced with this approach; thus, the maximal number of MUs are activated with optimal discharge frequency and the biomechanical parameters of movement and intermuscular coordination are similar to the analogous values in a main sport exercise. A trainee then "learns" to enhance and "memorize" these changes in motor coordination (evidently on a subconscious level).

We saw earlier that with this method the magnitude of resistance should be close to TF_{mm}. To avoid high emotional stress, CF_{mm} must be included only intermittently in the training routine. If the aim of a training drill is to "train movement" (i.e., both intramuscular and intermuscular coordination are the object of training), the recommended number of repetitions per set is one to three. Exercises such as the snatch or the clean and jerk are examples (see Figure 4.5 above). When the training of muscles rather than movement training is the drill objective (i.e., the biomechanical parameters of the exercise and intermuscular coordination are not of primary importance since the drill is not specific and is different in technique from the main exercise), the number of repetitions increases. One example is the "inverse curl" (Figure 4.6), where the typical number of repetitions is four to eight. The number of repetitions in squatting, on the other hand, usually falls in the range of two to six (Figure 4.2).

Although the method of maximum efforts is popular among superior athletes, it has several limitations and cannot be recommended for beginners. The primary limitation is the high risk of injury. Only after the proper technique for the exercise (e.g., barbell squat) is acquired and the relevant muscles (spine erectors and abdomen) are adequately developed is it permissible to lift maximal weights. In some exercises, such as sit-ups, this method is rarely used. Another limitation is that maximum effort, when employed with a small number of repetitions (one or two), has relatively little ability to induce muscle hypertrophy. This is the case because only a minor amount of mechanical work is performed and the amount of degraded contractile proteins is in turn limited.

Finally, because of the high motivational level needed to lift maximal weights, athletes using this method can easily become "burned out." The staleness syndrome is characterized by

- decreased vigor,
- elevated anxiety and depression,
- sensation of fatigue in the morning hours,
- increased perception of effort while lifting a fixed weight, and
- high blood pressure at rest.

This response is typical if CF_{mm} rather than TF_{mm} is used too frequently in workouts. Staleness depends not only on the weight lifted but also on the type of exercise used. It is easier to lift maximal weights in the bench press (where the barbell can be simply fixed and where the leg and trunk muscles are not activated) than in the clean and jerk, where demands for the activation of leg and trunk muscles, and for balance and arousal, are much higher.

Strength Training Methods

An athlete's best performance in a front barbell squat is 100 kg. He is able to lift this weight in a given set, one time only (one repetition maximum, or 1 RM).

The athlete has the following variants from which to choose for strength training:

- Lift 100 kg (maximal effort method);
- Lift a load smaller than 100 kg, perhaps 70 kg, either a submaximal number of times (submaximal effort method) or until failure (repeated effort method);
- Lift (move) a submaximal load at maximal velocity, for example, jump for height with a heavy waist belt (dynamic effort method).

Submaximal Effort Method and Repeated Effort Method

Methods using submaximal versus repeated efforts differ only in the number of repetitions per set—intermediate in the first case and maximal (to failure) in the second. The stimulation of muscle hypertrophy is similar for the two methods. According to the energetic hypothesis described in chapter 3 (muscle dimensions), two factors are of primary importance for inducing a discrepancy in the amount of degraded and newly synthesized proteins. These are the rate of protein degradation and the total value of performed mechanical work. If the number of lifts is not maximal, mechanical work diminishes somewhat. However, if the amount of work is relatively close to maximal values (e.g., if 10 lifts are performed instead of the 12 maximum possible), then the difference is not really crucial. It may be compensated for in various ways, for instance by shortening time intervals between sequential sets. It is a common belief that the maximal number of repetitions in a set is desirable, but not necessary, to induce muscle hypertrophy.

The situation is different, though, if the main objective of a heavy resistance drill is to "learn" a proper muscle coordination pattern. We can analyze this issue with an example. Suppose an athlete is lifting the barbell to 12 RM with a given rate of one lift per second. The muscle subjected to training consists of MUs having different endurance times from 1 to perhaps 100 s (in reality, some slow MUs have a much greater endurance time; they may be active for dozens of minutes without any sign of fatigue). The maximal number of lifts to fatigue among MUs varies, naturally, from 1 to 100. If the athlete lifts the barbell only one time, one division of the MUs is recruited and the second is not (Figure 4.7, left motoneuron column). According to the size principle, the slow,

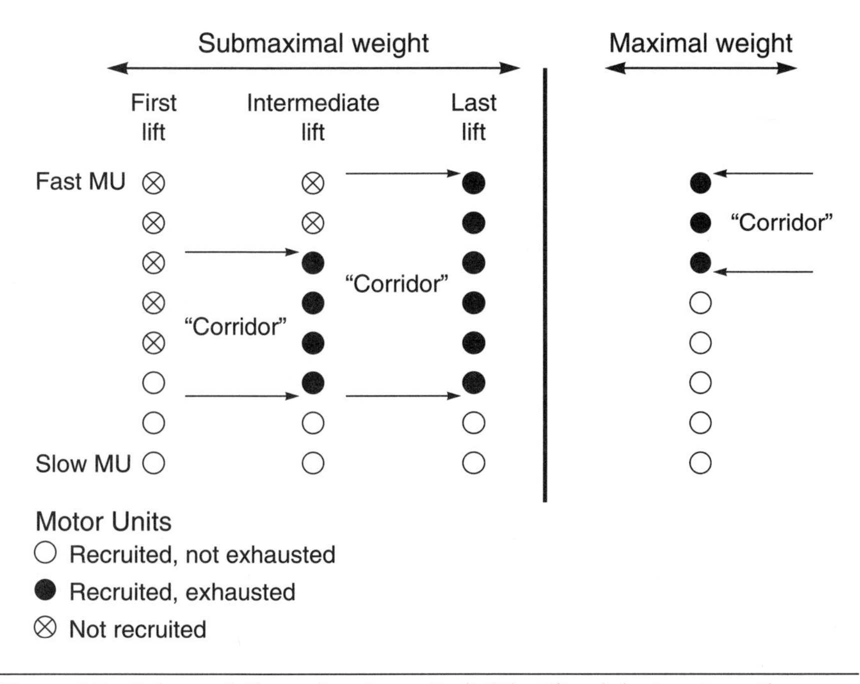

Figure 4.7 Subpopulations of motor units (MU) utilized during strength exercises.

fatigue-resistant MUs are recruited first (the slow MUs are shown in the lower portions of the MU columns). After several lifts, some of the recruited MUs become fatigued. Obviously, MUs possessing the shortest endurance time become exhausted. After six repetitions, for instance, only MUs with an endurance time under 6 s are exhausted. Since the exhausted MUs cannot now develop the same tension as at the beginning, new MUs are recruited. These newly recruited MUs are fast and nonresistant to fatigue. Thus they become exhausted very quickly. If only 10 lifts of the 12 maximum possible are performed, the entire population of MUs is distributed into three divisions (intermediate lift column in Figure 4.7).

1. **MUs that are recruited but not fatigued.** If they are not fatigued, they are not trained. All MUs having an endurance time above 10 s are in this category. It is evident that this subpopulation consists of slow MUs. The slow MUs are recruited at a low level of the required force and thus are activated regularly during everyday activities. Nevertheless, without special training their force does not increase. The conclusion that seems warranted from this finding is that it is very difficult to increase the maximal force of slow, fatigue-resistant MUs. Thus, a positive correlation exists between strength enhancement and the percentage of fast-twitch

muscle fibers. Individuals with a high percentage of fast MUs not only tend to be stronger but also gain strength faster as a result of strength training (Figure 4.8).

2. **MUs that are recruited and exhausted.** These are the only MUs subjected to a training stimulus in this set. These MUs possess intermediate features. In this subpopulation, there are no slowest MUs (recruited and fatigued) or fastest MUs (not recruited). The "corridor" of MUs subjected to a training stimulus may be relatively "narrow" or relatively "broad" depending on the weight lifted and the number of repetitions in a set. One objective of a strength program can be to increase the subpopulation of MUs influenced by training, or to broaden the corridor.

3. **MUs that are not recruited and therefore not trained.** If the exercise is performed to failure (repeated effort method), the picture is changed in the final lifts. A maximal number of available MUs are now recruited. All MUs are divided into two subpopulations: exhausted (fatigued) and nonexhausted (nonfatigued). The training effect is substantial on the first group only. If the total number of repetitions is below 12, all MUs with endurance

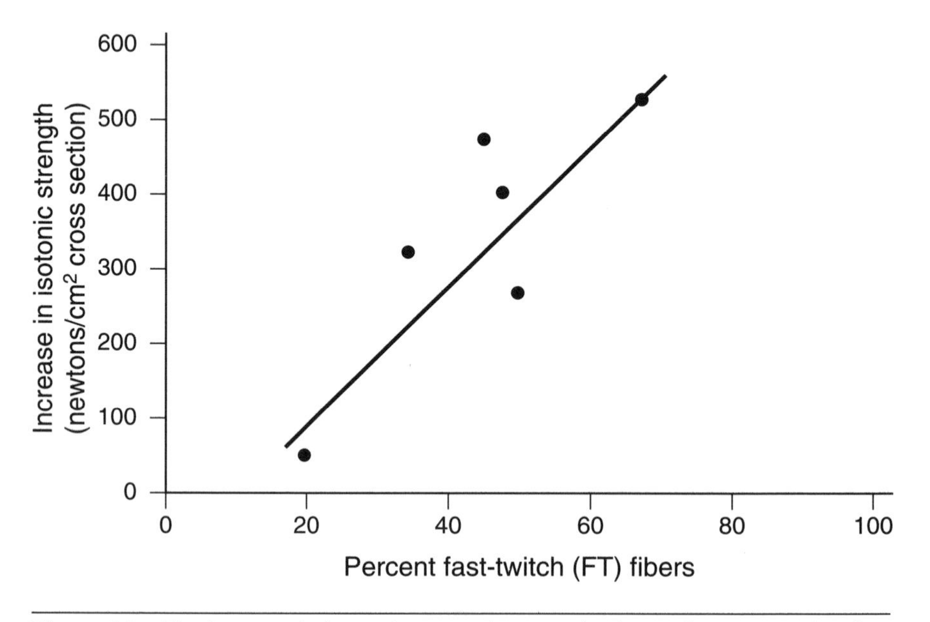

Figure 4.8 The increase in isotonic strength per unit of muscle cross-sectional area versus the percentage of fast-twitch muscle fiber distribution. Note. Adapted from "The Effects of Weight-lifting Exercise Related to Muscle Fiber Composition and Muscle Cross-Sectional Area in Humans" by B. Dons, K. Bollerup, F. Bonde-Petersen, and S. Hancke, 1979, *European Journal of Applied Physiology*, **40**, pp. 95-106.

times above 12 s fall into the second group. In spite of their early recruitment, these MUs are not exhausted (because of their high endurance).

When maximal weights are lifted (maximal effort method), the MU corridor includes a smaller number of MUs (see right column in Figure 4.7) than is the case when a submaximal weight is lifted a maximum possible number of repetitions. This is certainly a disadvantage for the method of maximal efforts. Only fast MUs are subjected to the training effect in this case. However, the advantages of the method outweigh any drawbacks.

To use the repeated effort method, the athlete must lift the weight with sincere exertions to failure (maximum number of times). This requirement is very important. Popular sayings among coaches—"Do it as many times as you can and *after that* three more times" and "No gain without pain"—reflect the demand very well. With this method, only final lifts in which a maximal number of MUs are recruited are actually useful. If an athlete can lift a barbell 12 times but lifts only 10, the exercise set is worthless.

Compared to the maximal effort method, the repeated effort method has pros and cons. There are three important advantages to the repeated effort approach. It has a greater influence on muscle metabolism and consequently on the inducement of muscle hypertrophy. In addition, it involves a greater subpopulation of trained MUs (the corridor; compare the two right columns in Figure 4.7) and poses a relatively low injury risk. This method also has limitations. The final lifts in a set are executed when the muscles are tired; thus, this training is less effective than lifting with maximal weights. Moreover, the very large training volume (the total amount of weight lifted) restricts the application of this method in the training of qualified athletes. Note, however, that the large amount of mechanical work performed can be considered an advantage if the objective of the exercise is general health and fitness rather than specific strength enhancement.

All the methods described are, and should be, used together in the strength training of qualified athletes. Many have attempted to answer the question: What kind of training is more effective—lifting of maximal or intermediate weights? This is similar to the question of whether 800-m runners should run in training distances shorter or longer than 800 m. They should run both. The same holds true for athletes training strength; they should employ exercises with different RMs.

Dynamic Effort Method

Because of the existence of the explosive strength deficit (see chapter 2; time), it is impossible to attain F_{mm} in fast movements against intermediate resistance. Therefore the method of dynamic effort is used not for increasing maximal strength but only to improve the rate of force development and explosive strength.

In conclusion, these combinations of methods can increase the maximum strength F_{mm}:

Method	*Immediate purpose*
Maximal efforts (use repeated efforts as a second choice)	Improve neuromuscular coordination • MU recruitment • Rate coding • MU synchronization • Coordination pattern
Repeated efforts (and submaximal efforts or both)	Stimulate muscle hypertrophy
Repeated efforts	Increase the corridor of recruited and trained MUs

Summary

Training intensity can be estimated by the

- magnitude of resistance (i.e., weight lifted) expressed as a percentage of the best achievement attained during a competition (CF_{mm}) or in training (TF_{mm});
- the number of repetitions (lifts) per set; and
- the number (or percentage) of repetitions with maximal resistance (weight).

Exercising at varying levels of resistance causes differences in (a) metabolic reactions, (b) intramuscular coordination, and (c) biomechanical variables and intermuscular coordination. The produced mechanical work as well as the metabolic energy expenditures increase as the weight lifted decreases.

The total amount of degraded protein, however, is the function of both the mechanical work performed (or the total weight lifted) and the rate of protein catabolism. According to the energetic hypothesis, the rate of protein degradation is a function of the weight lifted: The heavier the weight, the higher the rate of protein degradation. The mass of proteins catabolized during heavy resistive exercise is a product of the rate of protein breakdown and the number of lifts. The mass is maximized when training intensity is between 5 to 6 and 10 to 12 RM.

When an athlete lifts maximal weight, a maximum number of MUs are activated; the fastest MUs are recruited; the discharge frequency of motoneurons is at its highest; and MU activity is synchronous. One objective of heavy resistance training is to "teach" an athlete to recruit all the necessary MUs at a firing rate that is optimal for producing a fused tetanus in each motor fiber.

Elite weight lifters use a broad spectrum of loads, but the largest proportion of weights lifted is composed of those 70% to 80% of the CF_{mm}. The average weight these athletes lift is 75.0 ± 2.0% of CF_{mm}. These repetition levels should not be mechanically copied, but rather, thoughtfully implemented.

Strength training can be accomplished in three ways: (a) lifting a maximum load (exercising against maximal resistance)—the maximal effort method; (b) lifting a nonmaximal load to failure, with the muscles developing the maximum force possible in a fatigued state during the final repetitions—the repeated effort method; and (c) lifting (or throwing) a nonmaximal load with the highest attainable speed—the dynamic effort method. In addition, the lifting of nonmaximal loads an intermediate number of times (not to failure) is used as a supplementary training method (the submaximal effort method).

The maximal effort approach is considered superior for improving both intramuscular and intermuscular coordination: The maximal number of MUs is activated with optimal discharge frequency. When you use this training method, the magnitude of resistance should be close to TF_{mm}. If the aim of a training drill is to "train movement," the recommended number of repetitions per set is one to three. When the aim is to "train muscles," on the other hand, the number of repetitions increases. The maximal effort method, while popular among superior athletes, has several limitations (such as the high risk of injury, staleness). It also has a relatively small potential to stimulate muscle hypertrophy.

The submaximal effort and the repeated effort methods are similar in their ability to induce muscle hypertrophy. They are, however, rather different with respect to training muscular strength, especially improving the neuromuscular coordination required for maximal force production. The submaximal effort method (the lifting of nonmaximal loads, but not to failure) does not seem to be effective for enhancing muscular strength and improving specific intramuscular coordination. With the repeated effort approach, the weight must be lifted to failure: Only final lifts, in which a maximal number of MUs are recruited, are actually useful. The saying, "No pain, no gain" reflects this demand.

CHAPTER 5

Timing in Strength Training

The distribution of exercises and a training load over certain time periods (i.e., the *timing of training*) is a matter of primary importance for the outcome of an athlete's preparation. Two main problems in this area are how to space work and rest intervals and how to sequence exercises.

Structural Units of Training

Training can be divided into structural units. It is customary to identify these structural units as training session (workout), training day, microcycle,

Table 5.1 Everyday Training Schedule of the Bulgarian Olympic Weight-Lifting Team

Time	Mon., Wed., Fri.	Tues., Thurs., Sat.
9:00–10:00	Snatch	Snatch
10:00–10:30	Rest	Rest
10:30–11:30	Clean and jerk	Clean and jerk
11:30–12:30	Exercise	Exercise
12:30–1:00	Rest	Exercise
1:00–5:00	Rest	Rest
5:00–5:30	Exercise	Exercise
5:30–6:00	Exercise	Rest
6:00–6:30	Rest	Exercise
6:30–8:30	Exercise	Rest
Total Exercise Time	6 hr	4.5 hr

mesocycle, macrocycle, Olympic cycle (quadrennial cycle), and long-lasting, or multiyear, training.

The *training session (workout)* is generally viewed as a lesson comprising rest periods not longer than 30 min. The reason for such a definition, which initially appears too formal, is the need to describe training in sports in which a daily portion of exercises is distributed among several workouts. An example of such a training routine is given in Table 5.1. According to the definition, athletes in this instance have only two workouts a day. Training drills separated by 30-min rest intervals are considered part of one training session. This all-day schedule is a good example of the *training day* of world-class athletes. One renowned athlete once joked, "My life is very rich and diversified. It consists of five parts—training, competitions, flights, sleep, and meals." This is very close to reality.

To appraise the training load of different workouts, the time needed to recover from one training session is used, according to the following classification:

Training load of one workout	*Restoration time, hr*
Extreme	> 72
Large	48–72
Substantial	24–48
Medium	12–24
Small	< 12

A *microcycle*, the third category, is the grouping of several training days. The run of a microcycle in the preparation period is usually 1 week. In the competition period, the length of a microcycle is typically adjusted to the duration of the main competition. For instance, if a competition in wrestling lasts 3 days, it is advisable to employ microcycles of the same duration. Usually, a general framework of microcycles is routinely reproduced over a relatively long period of time (that is why it is called a cycle).

A *mesocycle* is a system of several microcycles. Typically, the duration is 4 weeks with a possible range of 2 to 6 weeks. The duration and even the existence of mesocycles in the training of eastern European athletes were influenced by the practice of "centralized preparation." The best athletes were once prepared in training camps throughout the year, mainly for logistical reasons; shortages in food and other important goods at home made it impossible to create normal training conditions. Such training management has its pros and cons, however. The enhanced competitiveness and increased possibilities for obtaining and sharing information are positive features. On the other hand, the standard environment, life without one's family, and the necessity of living and communicating with a single group of people, who are often rivals in the same sport, impose additional psychological burdens on the athlete. To reduce this psychological stress and to diversify the environment, training camp locations were regularly changed. Interviews of the athletes showed that they preferred 4-week training camps interspersed with 1- or 2-week visits home.

There is no reason to reproduce in full this pattern in the West; however, some elements of the described training timing, mesocycles included, are undoubtedly useful. *Mesocycles* may be classified according to the objective of training as accumulative, transmutative, and realizational. The aim of *accumulative mesocycles* is to enhance the athlete's potential, that is, to improve basic motor abilities (conditioning) as well as sport technique (motor learning). The outcome of an accumulative mesocycle is evaluated on the basis of tests (e.g., measures of strength or aerobic capacity), the athlete's performance in auxiliary exercises, and the quality of technical skill. In these mesocycles, various exercises, including nonspecific ones, may be used for conditioning.

The *transmutative mesocycles* are employed to transform the increased nonspecific fitness into specific athlete preparedness. Throughout this period, specific exercises are mainly used for conditioning and polishing sport technique. The performance during unofficial or nonimportant contests is used primarily to estimate training progress. The *realizational*, or *precompetitive, mesocycles* are planned to put on the best sport performance attainable within a given range of fitness. Performance during an important competition is the only measure of success or failure during this period.

At the next structural level, *macrocycle* refers to one entire competition season and includes preparation, competition, and transition periods (phases). Each period consists of several mesocycles. The typical duration of a macrocycle is one year (for winter sports) or half a year (for track and field events in which both indoor and outdoor competitions are held). In wrestling and swimming, there are three macrocycles in a year. The organization of training programs into macrocycles and periods of training is called *periodization*. Still more long-ranged views are helpful as well.

The Olympic cycle is quadrennial, four years in length, from one Olympic games to another. And longstanding, or multiyear, training embraces the career of the athlete, from beginning to end.

Planning workouts, training days, microcycles, and mesocycles comprise short-term planning. Planning macrocycles is called medium-term planning. Long-term planning deals with training intervals of many years.

Short-Term Planning

In short-term planning, the effects of fatigue are the primary influencing factor. For instance, a training session should be designed so that exercises (such as strength, speed, or technique exercises) directed toward improving fine motor coordination (central factors), rather than peripheral factors, are performed in a fresh, nonfatigued state, preferably immediately after warm-up. In endurance sports, however, when the aim is to improve velocity at the finish of a distance rather than the maximal speed attainable in a fresh state, speed exercises may be performed after endurance work.

Paradigm of Timing Short-Term Training

A general principle of short-term training design is that *fatigue effects from different types of muscular work are specific*. This means that an athlete who is too tired to repeat the same exercise in an acceptable manner may still be able to perform another exercise to satisfaction. Changing or sequencing a drill appropriately makes it possible to assign more labor and to suitably increase the training load. For instance, if a trainee performs a leg exercise, such as squatting, and an arm exercise, such as bench press, the total number of lifts will be greater when the exercise sequence is bench press, squatting, bench press, squatting, and so on than when the sequence is squatting, squatting, bench press, bench press. The same principle is valid for exercises of different directions, for instance, strength and sprint exercises. The fatigue effect from a heavy resistance exercise routine mainly affects the possibility of performing or repeating an exercise of this type. Thus, one's ability to execute drills of another type is restored more quickly than one's ability to repeat the same routine (Figure 5.1).

You will find that if two similar training workouts are executed in a row, the traces of fatigue from the two sessions are superimposed (Figure 5.2). If the training load is large (i.e., the restoration time takes from 48 to 72 hr), several training sessions of this type performed sequentially may lead to severe exhaustion of the athlete.

If exercises with different targets could be trained all the time, it would be easy to distribute these exercises among training sessions to avoid the superimposition of fatigue traces. However, fitness gain decreases if several motor abilities are trained simultaneously during one workout, microcycle, or mesocycle. Therefore, it is not a good idea to have more than two or three main targets in one micro- or mesocycle. For instance, there is no reason to train, in one microcycle, maximal strength, explosive strength, aerobic capacity, anaerobic lactacid and alactacid capacities, maximal speed, flexibility, and sport technique. The organism cannot adapt to so many different requirements at the same time. The gains in all these motor abilities would be insignificant compared with the gain from development of only one physical quality. When the training targets are distributed over several mesocycles in sequence, the fitness gain increases.

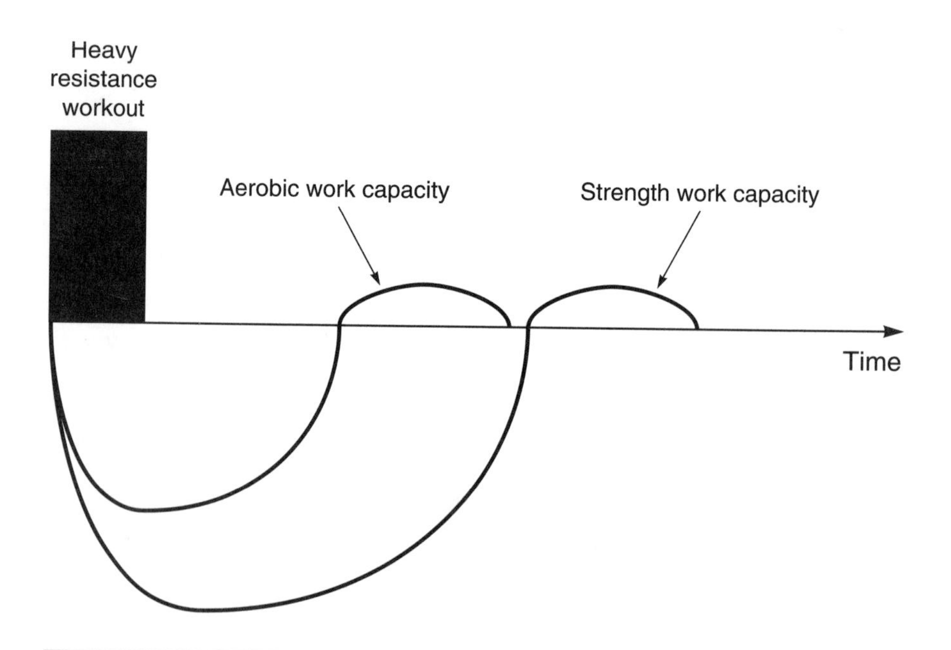

Figure 5.1 The time course of athlete restoration after a heavy resistance training session.

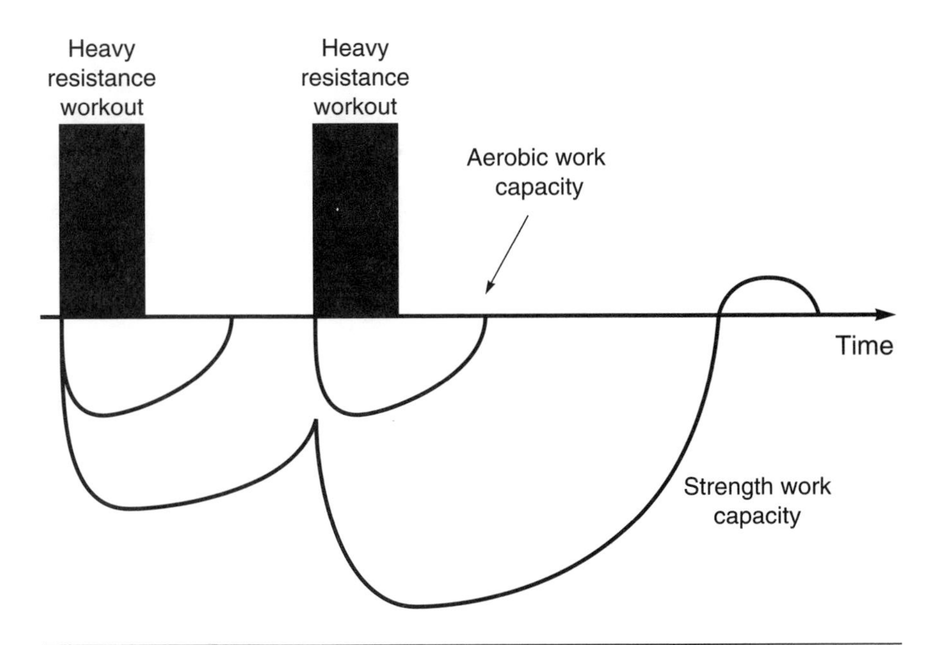

Figure 5.2 The superposition of two resistance workouts and their effects.

Clearly, a conflict exists between the tendency to decrease the number of training targets and the tendency to increase the range of targets in a training program. Coaches or athletes decrease the number of targets in micro- and mesocycles—in other words, use *specialty programs*—to enhance performance growth.

In contrast, they increase this number, using *combined programs*, in order to have more freedom in planning the training schedule to avoid the superimposition of fatigue traces from individual workouts and the hazard of staleness.

Similar contradictions are, in general, typical for the planning of various training programs. The problem is to find a proper balance between the conflicting demands. Some world-class athletes have found that two is the optimal number of motor abilities or targets that can be improved in one mesocycle. In addition, only one essential feature of sport technique (such as tolerance to fatigue, stability) can be trained within this interval. Up to 70% to 80% of total work within the mesocycle should be addressed to the development of the targeted motor abilities (about 35%-40% per target).

Workouts and Training Days

The general idea in planning strength training sessions is to have the athlete do as much work as possible while being as fresh as possible. Unlike the situation for endurance training, it is not necessary for the athlete to become exhausted in a heavy resistance *workout* (do not confuse this with exercise *set*). Strength gains are greater if trainees exercise when they are not tired. This is especially true when the target of the resistance exercise is neural coordination, both intra- and intermuscular. Broadly speaking, a trainee should "learn" to either decrease inhibitory output or enhance excitatory output from the central nervous system (CNS) while exercising and thereby gain strength. This learning is more successful if the trainee is fully recovered from previous activity, not fatigued. To have athletes exercise while they are as fresh as possible, the training workouts should be carefully planned.

The timing of workouts has three facets: rest-exercise alternation, exercise sequence, and intensity variation.

Rest-Exercise Alternation

In general, the large interbout rest intervals are usually employed in heavy resistance training aimed at increasing muscular strength.

The total number of sets per day has not changed in the preparation of elite weight lifters over the last 40 years (most average 32–45 sets, but some athletes manage 50–52). However, the duration of a workout has changed; in 1955–1956 it was only 2 to 2.5 hr, and in 1963–1964 it was 3 to 3.5 hr (one training workout a day was used). Since 1970, two and more training sessions a day have been the rule. The same number of sets is distributed now among two or more daily workouts.

Bulgarian athletes, for instance, have several workouts a day with a total duration of up to 6 hr (see Table 5.1). The exercise sessions are limited to 60-min, or even 45-min, periods. Two sessions in the morning and two in the afternoon are separated by 30-min rest intervals. The underlying assumption is that the elevated blood testosterone level can be maintained for 45 to 60 min only and that a 30-min rest is needed to restore the testosterone level. (This assumption has not been proven; the precise nature of the elevated testosterone level during a strength exercise workout is not well understood. In general, the elevation may be induced either by increased testosterone production or by a decreased amount of testosterone acceptors in muscles and other tissues.) During the 30-min rest intervals the athletes may choose to lie down and listen to music. To avoid cooldown, they are warmly dressed; their relaxed legs are slightly raised, supported by a small bench.

Both sport practice and scientific investigations have demonstrated that distribution of the training volume into smaller units produces effective adaptation stimuli, especially for the nervous system, provided that the time intervals between workouts are sufficient for restitution.

To prevent early fatigue, rest intervals between sets by elite athletes, especially when working with large weights, are approximately 4 to

5 min. Since the duration of a strength exercise bout is short, the *exercise/rest ratio* (i.e., the duration of the bout relative to rest) is very small for this type of physical activity. However, even long rest periods of 4 to 5 min are not sufficient for complete recovery, which after the lifting of a *maximum training weight* (TF_{mm}) requires 10 to 15 min. If the duration of a strength training workout is limited, one possible solution is to combine sets into series and schedule long (10–12 min) rest intervals between them. Since working periods are short and rest periods are long during sessions, *workout density* (the number of sets per hour of a workout) is not considered an informative measure of strength training intensity.

Exercise Sequence

The idea in sequencing exercises is to perform the most valuable exercises requiring fine motor coordination and maximal neuronal output in a rested state. To avoid premature fatigue, which is detrimental to a subsequent exercise, the following advice is usually given:

- Include main sport exercises before assistance exercises.
- Use dynamic, power-type drills before slow exercises, such as squats.
- Exercise larger muscle groups before smaller ones.

If the target of a workout is to increase muscle strength (not induce muscle hypertrophy; see chapter 8), successive exercises should minimally involve the same muscle groups. A sequence such as (a) arm abduction with dumbbells (only deltoid muscles are active), (b) bench press (same muscles are involved), (c) front squat (assistance exercise, performed with relatively slow speed), and (d) snatch (competition lift; requires maximal power production and complex technique) would prove incorrect. The proper series would look like this:

1. Snatch
2. Bench press
3. Squat
4. Arm abduction

Intensity Variation

Because lifting a maximum training weight (maximum effort method) is recognized as the most efficient way to train, this should be practiced at the beginning of a training workout, following the warm-up. Then athletes perform a few (two or three) single lifts toward the training weight expected for the date and several (up to six) sets with this weight. Bulgarian athletes use a trial-and-error approach to achieve TF_{mm} every day. Russian coaches typically plan the exercise intensity in advance, considering a load 90% of CF_{mm} as TF_{mm}. A complex of combined exercise sets, for instance in snatch lifts, lasts a maximum of 30 min (6 sets × 5 min for rest intervals).

"Pyramid" training, popular many years ago, involves gradually changing the load in a series of sets in an ascending and then a descending manner. This has been virtually abandoned by Olympic-caliber athletes. The ascending part of such a routine induces premature fatigue, while the descending portion is not efficient since it is performed in a fatigued state. For contemporary training, fast progression to the main training load is typical.

■ *Past and Present*

From the training log of the Olympic (1960) champion Victor Bushuev. Drills in military (standing) press.

> *Year 1958.* TF_{mm} was 90 kg. The conventional [at that time] "pyramid" protocol was executed. The following weights were lifted: 60, 65, 70, 75, 80, 85, 90, 90, 85, 80, 75, 70 kg. Each weight, except in the beginning sets utilizing 60- and 65-kg barbells, was lifted until failure. The initial part of this protocol (in the range 60–85 kg), which was not very useful, induced substantial fatigue and decreased the effect of lifting the highest loads.
>
> *Year 1960.* TF_{mm} was 110 kg. The barbell weight varied in the following sequence: 70 kg, 90 kg, 100 kg (all three weights were lifted only one time in a set), and then 110 kg. The maximal weight was lifted in five sets, one to two times in each set.

Since 1964, "pyramids" have been virtually excluded from the training of elite strength athletes.

A couple of other points about intensity variation are useful in specific circumstances. Athletes who are feeling fatigue may take single lifts at 10 to 15 kg below TF_{mm} between maximum lifts. This is also helpful for the purpose of recalling a proper technique pattern. Finally, if both the maximum effort and the repeated effort methods are used in the same workout, maximal lifts should be included first.

Advice about exercise sequence and intensity variation may be extended to the planning of a training day. Thus, exercises requiring maximal neural output (e.g., competitive lifts, power drills, lifting TF_{mm} or CF_{mm}) should be performed in a fresh state when the athlete has recovered from previous activity (i.e., during morning training workouts).

Contrasting Exercises

It is advisable to schedule flexibility and relaxation exercises between heavy resistance drills to speed up recovery and prevent loss of flexibility. The preferred area for flexibility exercises is the shoulder joints.

Mixed Training Sessions

Sessions that include the strength routine as a section are less effective than special heavy resistance workouts. In sports in which muscular strength is the ability of primary importance (e.g., field events in track and field, American football), it is especially advisable to set apart heavy resistance drills in a separate workout. If there is not enough time to do this, strength exercises may be included in mixed workouts. To prevent negative effects, they are usually performed at the end of workouts (this practice is accepted in gymnastics, and other sports). However, a coach should be aware that the same strength training complex is more effective when used at the beginning of the training session when the muscles are not fatigued.

■ *Are Special Strength Training Sessions Necessary?*

Muscular strength is only one of several abilities athletes must utilize; they have many other things to develop besides strength. It is up to the coach to decide whether or not to spend time on special strength training sessions. In many sports, such as tennis and even men's gymnastics, it is possible to attain the required level of strength fitness by performing strength exercises immediately after main workout drills. However, if low strength levels actually limit athletic performance, special strength workouts are useful. The junior soviet/Russian team in men's gymnastics employs separate strength workouts; the men's team does not. In many sports, such as track and field, rowing, and kayaking, heavy resistance workouts are part of the routine. In others, such as swimming and wrestling, workouts are for specific strength exercises rather than for heavy resistance training (dryland training in swimming, imitation of takedowns with simulated or added resistance in wrestling).

Circuit Training

The idea of circuit training is to train several motor abilities (especially strength and endurance) at the same time. Such programs consist of several (up to 10 or 12) stations with a given exercise to be performed at each one.

The basic philosophy of circuit training (to stimulate strength and endurance simultaneously) appears dubious. It is well known that the mechanisms of biological adaptation to strength and endurance types of physical activity are different (this issue will be discussed in chapter 8). The muscles are not able to optimally adapt to both types of exercise at the same time. Combining strength and endurance exercises interferes with the ability to gain strength. Conversely, vigorous endurance activity inhibits strength development (Figure 5.3).

Because of the low strength gains (in comparison with those obtained from regular strength training routines), circuit training is not recommended and is hardly ever used in strength and power sports. It may, however, be employed in sports having a high demand for both strength and endurance (rowing, kayaking) and also for conditioning in sports in which strength is not the dominating motor ability (volleyball, tennis). Circuit training is mainly used by athletes primarily concerned with enhancing or maintaining "general fitness" rather than specific muscular strength.

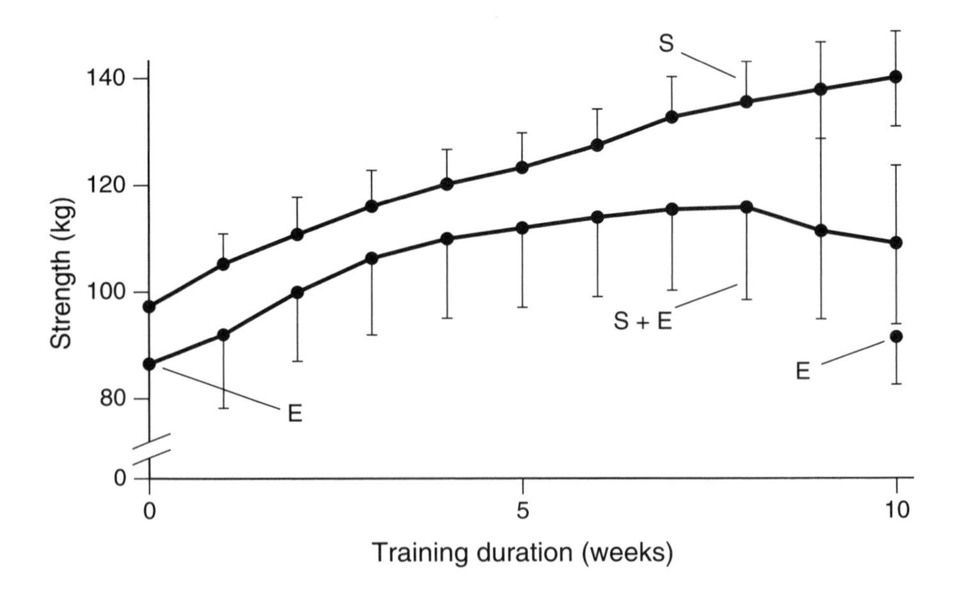

Figure 5.3 Simultaneous training for endurance (E) and strength (S) inhibits strength gain. Note. From "Interference of Strength Development by Simultaneously Training for Strength and Endurance" by R.C. Hickson, 1980, *European Journal of Applied Physiology*, **45**, pp. 255-263.

Microcycles and Mesocycles

The timing of heavy resistance protocols in micro- and mesocycles is dominated by two main ideas. One is to allow adequate recovery between exercise periods, and the other is to find a proper balance between the steadiness of a training stimulus (to call forth an adaptation) and its variability (to avoid premature accommodation and staleness).

Adequate Recovery

During a microcycle, rest-exercise alternation and proper exercise sequencing alleviate fatigue. The greatest training adaptation to a standard stimulus occurs when muscles are recovered from previous training periods and best prepared to tolerate the greatest overload. Keeping in mind that qualified athletes have 5 to 6 training days a week, one may conclude that the restoration time after a workout should be about 24 hr—that athletes should train with small (restoration time less than 12 hr) and medium (12–24 hr) training loads. In this case, however, the total training load is not great enough to stimulate strength development. The solution is the proper exercise alternation in consecutive workouts. Since fatigue effects from different resistance exercises are specific, it is possible to increase training loads up to an optimal level by properly rotating the exercises in sequential sessions. Exercises in consecutive training sessions should minimally involve the same muscle groups and thus repeat the same pattern of muscle coordination. It would not make sense, for instance, to plan two consecutive workouts with the snatch lift.

Recovery time from heavy resistance exercises varies with muscle size. For small muscle groups, like the calf muscles, the restitution time is typically less than 12 hr. (Remember that we are concerned here with the training of experienced athletes only.) Small muscle groups (such as ankle plantar flexors and muscles of the forearm) may be trained several times a day. Intermediate muscle groups require more time for restitution; these can be exercised every day. Finally, it is advisable to exercise the large muscle groups with rest periods of at least 48 hr. For instance, barbell squats are performed usually only two times a week with 72 or 96 hr between training sessions (Olympic-caliber weight lifters perform front and back squats up to two times per week). The squats are excluded from training programs 1 week (in weight lifting) or 10 to 12 days (track and field) before an important competition begins.

To increase muscle strength, the schedule should include at least three heavy resistance workouts per week. It is better to distribute the same training volume into several workouts than to concentrate it into a small number of sessions. Athletes who increase the number of training sessions per week, while keeping the training volume (number of repetitions, total weight lifted) constant, usually experience visible strength gains. For instance, when the volume is distributed into two daily training sessions, the strength development is greater than it is with one session a day.

To retain strength gains, at least two training sessions per week should be scheduled.

Variability

The *variability* of training programs during micro- and mesocycles is realized through changes in training load (not exercise complexes). One stable complex of exercises should be performed through a mesocycle (to elicit an adaptation). This complex, consisting of perhaps 10 exercises, is distributed among the training days and workouts of one microcycle provided that each exercise is performed at least twice a week. The time order of exercises is kept constant from one microcycle to another. For example, the snatch and front squat are routinely performed during the morning workout of the first day of each microcycle.

To avoid premature accommodation, training loads should vary from day to day and from microcycle to microcycle. The empirical "rule of 60%" has stood the test of time: The training volume of a day (microcycle) with minimal loading should be around 60% of the volume of a maximal day (microcycle) load.

Stress (Impact) Microcycles

Some experienced athletes use stress microcycles, in which fatigue is accumulated from day to day (due to high training loads and short rest intervals that are insufficient for restitution), if a routine training program does not bring about strength gains. The microcycle after a stress microcycle should involve small training loads. Elite athletes may tolerate up to two stress microcycles in a row (*doubled stress microcycle*); however, coaches and athletes should exercise extreme caution with this approach. Stress microcycles should not be used more than three to four times per year. Doubled stress microcycles are used only once a year.

The *training volume* per 4-week mesocycle is approximately 1,700 lifts for elite (soviet/Russian) weight lifters; 1,306 ± 99 repetitions for qualified athletes having a Master of Sport title; and 986 ± 99 lifts for athletes with 1 year of experience in weight-lifting training.

Medium-Term Planning (Periodization)

The term *periodization* refers to a division of the training season, typically 1 year long, into smaller and more manageable intervals (periods of training, mesocycles, and microcycles) with the ultimate goal of reaching the best performance results during the primary competition(s) of the season. To do this the athlete changes exercises, loads, and methods during preseason and in-season training. When the same training routine is applied over the entire season from the early preparatory phase (preseason) to the in-season training, improvement occurs only in the early phase and there is a subsequent leveling off. Early staleness is almost unavoidable with such a protocol.

The Issue of Periodization

Periodization is regarded as one of the most complex problems in athlete training. The proper balance between opposing demands is difficult to achieve in medium-term planning because so many factors are involved.

The efficacy of planning in macrocycles is determined for groups of athletes, not for individuals, and is calculated as follows:

$$\text{Efficacy coefficient, \%} = \frac{100 \cdot \text{(The number of athletes who achieved their best performances during most important competition of the season)}}{\text{The total number of athletes}}$$

For national Olympic teams, such competitions as the Olympic games and world championships are regarded as the most important. An efficacy coefficient of about 85% is considered excellent, 75% is considered good, and 65% is considered acceptable.

In *medium-term planning*, four issues have primary importance:

1. *Delayed transformation* of training loads (into fitness development)
2. *Delayed transmutation* of nonspecific fitness acquired in assistance exercises (relative to a main sport skill) to a specific fitness
3. *Training residuals*
4. *The superposition of training effects*

Delayed Transformation

To conceptualize delayed transformation, imagine a group of athletes trained in the following manner. They perform the same exercise (e.g., dead lift) with a constant intensity (2–5 RM) and volume (five sets) during each workout (three times a week). At the beginning, maximal strength increases relatively fast; however, after 2 to 3 months of this standard training, the rate of strength enhancement decreases as a result of accommodation. To overcome the accommodation, the coach decides to increase the training load (the number of training sessions per week, sets in workouts). But after several weeks, the performance fails to improve again. This time the coach decides to decrease the training load. After a certain period, the athletic performance again begins to improve. This period is called the *period of delayed transformation* (of the training work into performance growth).

In general, during periods of strenuous training, athletes cannot achieve the best performance results for two main reasons. First, it takes time to adapt to the training stimulus. Second, hard training work induces fatigue that accumulates over time. So a period of relatively easy exercise is needed to realize the effect of the previous hard training sessions—to reveal the delayed training effect. Adaptation occurs mainly when a retaining or detraining load is used after a stimulating load.

The time of delayed transformation lengthens as the total training load and accumulated fatigue increase. Typically, the delayed transformation lasts from 2 to 6 weeks with the average time of 4 weeks—exactly one mesocycle. This mesocycle, we recall, is known as the realization, or pre-competition, mesocycle. Its objective is to prepare the athlete for immediate competition. The training load is low at this time. The main training work has been performed in preceding mesocycles (accumulation and transmutation). Because the effects are delayed, the adaptation occurs (or is manifested) during unloading rather than loading periods.

Delayed Transmutation

To continue the above example, when the athletes' achievements stop improving, the coach modifies the strategy and decides to change the exercises rather than the training load. Now, instead of performing the dead lift (which was the final training activity), the athletes begin to perform several assistance exercises such as leg and spine extensions and arm curls. After a couple of months of this training, the athletes' performances improve in all the drills except perhaps the only one—the dead lift—that was not trained. The athletes' potentials are now better than before; however, performance results in the dead lift are the same.

Now a special training routine must be advanced to transmute the acquired motor potential into athletic performance. Both special efforts and time are needed to attain this goal, which is realized during the *transmutation mesocycles*. Training in such mesocycles is highly specific. The number and total duration of transmutation mesocycles in one season depend on the total duration of preceding accumulation mesocycles. Transmutation and realization mesocycles, when considered together as one unit, are often called the *tapering* (or *peaking*) period.

Analysis of our example shows that both the training content (exercises used) and the training load should vary over an entire training season. The accumulation, transmutation, and realization (tapering) mesocycles follow one another in a certain order. To effectively plan these mesocycles (their duration, content, and training load), the coach or athlete must take into account training residuals and the superposition of training effects.

Training Residuals

The reduction or cessation of training brings about substantial losses in adaptation effects. However, athletes to a certain extent can sustain the acquired training benefits over time without extensively training them continually. De-adaptation, as well as adaptation, takes time. If athletes exclude a given group of exercises (e.g., maximal strength load) from training protocols, they gradually lose the adaptations. A positive correlation exists between the time spent to elicit adaptational effects, on the one hand, and the time of detraining, on the other (Figure 5.4).

Four factors mainly determine the time course of detraining: (1) duration of the immediately preceding period of training (the period of accumulation), (2) training experience of the athletes, (3) targeted motor abilities, and (4) amount of specific training loads during detraining (or retaining) mesocycles.

1. The general rule is that the longer the period of training, the longer the period of detraining, or "Soon ripe, soon rotten." When a preparatory period (preseason) is long, for instance several months, and a competition period is short (several weeks), as in many Olympic sports, it is permissible

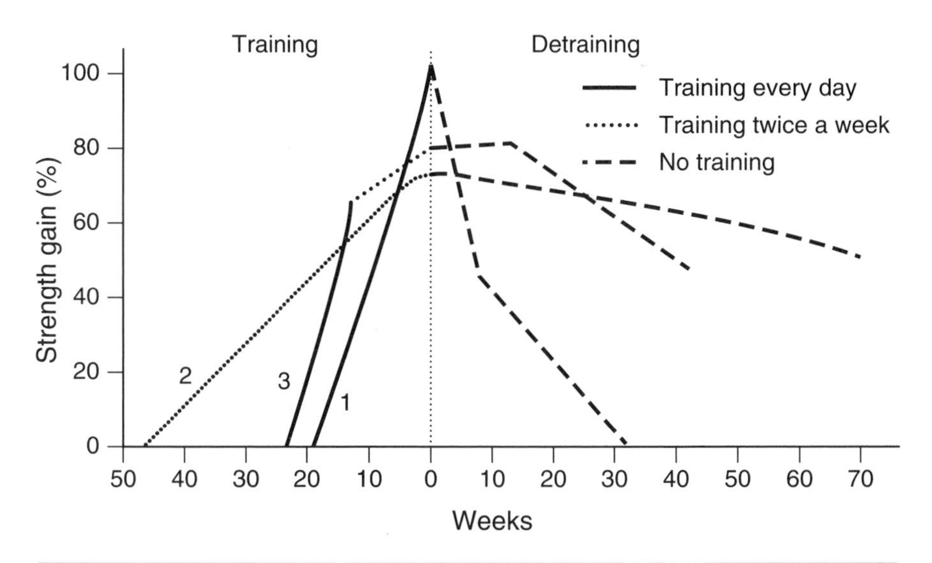

Figure 5.4 Time of training and detraining in three groups of subjects. Group 1 trained daily; group 2 had two training sessions a week; group 3 trained daily at the beginning and then twice a week. Note. From *Isometrische Muskeltraining* by T. Hettinger, 1966, Stuttgart: Fischer Verlag.

to eliminate certain exercises (like heavy resistance training) during the competition period. Strength adaptation is not lost in this case, mainly because the detraining period is short. However, in sports with a brief preparatory period and many competitions in a row (as with games in ice hockey or tournaments in tennis), strength gains elicited during the short preparatory period (weeks) are lost almost completely during the period of competition (months) if maximal strength loads are not used.

2. Mature athletes with continuous and extensive training backgrounds find that the residual effects of training are relatively stable. These athletes have slow rates of detraining and are able to achieve good results after relatively short periods of retraining. This is a result of both what they have done in the past and what they are presently accomplishing. Elite athletes with training backgrounds that span many years regain motor abilities much more quickly than average athletes.

3. Once special training ceases, different training benefits are lost at various rates. Anaerobic capacities are lost very quickly whereas adaptations to aerobic or maximal strength loads are relatively long lasting. The most stable benefits are training residuals based on morphological changes in skeletal muscles. Muscle size, for instance, changes slowly during both training and detraining. Because of this it is possible to use specialty mesocycles in which motor abilities are developed in sequence. The attained level of the motor ability (e.g., maximal strength) that was the primary target in one mesocycle is maintained in subsequent mesocycles with small loads.

4. If special training loads (e.g., heavy resistance exercises) are preserved at a certain level, it is possible to either retain the acquired level of the specific motor ability or lose it at a relatively low rate. A coach may prescribe specific retaining or detraining loads for a given period during which training residuals are conserved at appropriate levels (but not improved).

The Superposition of Training Effects

Various training methods do not bring the same gross beneficial effect to all physiological systems. Training effects are specialized and they affect separate systems in different ways. Methods that induce beneficial adaptation in one motor ability or physiological system may produce negative effects on another ability or system. For example, excessive strength gains associated with muscle hypertrophy may have negative effects on aerobic endurance as a result of reduced capillary density in the working muscles.

The transfer of either positive or negative effects between two types of training is not necessarily symmetrical. In other words, the effect of training activity *X* on ability *Y* is often different from the effect of training activity *Y* on ability *X*. Usually, hard strength training affects aerobic endurance negatively. The counterproductive effect of aerobic endurance training on maximal strength, if it exists, is smaller. Because of this, the strength-aerobic endurance sequence in two consecutive mesocycles provides a definite advantage; strength gains achieved during the first mesocycle are not minimized by aerobic training during the second. The opposite sequence, aerobic endurance-strength, is less efficient. In this sequence, aerobic capacity is initially enhanced but then deteriorates during the ensuing mesocycle.

Periodization as a Trade-Off

Though most people understand the necessity of varying both training loads and training content over an entire season, being able to prescribe the optimal training plan for a given athlete and predict its effect on sport performance is not easy. This area of planning is contentious. In reality, a good periodization plan is a subtle trade-off between conflicting demands.

On the one hand, an athlete cannot develop maximum strength, anaerobic endurance, and aerobic endurance all at the same time. The greatest gains in any one direction (for instance, strength training or aerobic training) can be achieved only if an athlete concentrates on this type of training for a reasonably long time—at least one or two mesocycles. In this case, the improvement in strength or aerobic capacity will be more substantial than that achieved with a more varied program. This leads to the recommendation that *one should train sequentially—one target after the other*. Elite athletes have favored this widely used approach for many years (Figure 5.5a).

In the 1960s, for instance, middle-distance runners used a preparation period of 7 months consisting of the following sequence: (a) aerobic training, called at that time "marathon training" or "road training"—2.5 to 3 months; (b) "hill training" or uphill running, mostly anaerobic with an increased resistance component—2.5 months; and (c) "short-track" training in a stadium—about 1.5 months. This training plan corresponds to the saying that athletes should begin a workout from the short end and a season from the long one. Similarly, throwers began the preparation period with strength exercises and, only after 2.5 to 3 months of such training, began more specific routines.

But with this approach, because of the great amount of time and effort spent in a specific direction, an athlete has little opportunity to perform other drills or exercises. As a result of the long periods of detraining, the level of nontargeted motor abilities decreases substantially. In addition, great physical potential (for instance, strength, aerobic capacity) acquired in periods of accentuated training does not directly involve the sport movement. That is, the strength level is improved (for instance, dryland training in swimmers), but athletic results are not. Much time and effort is

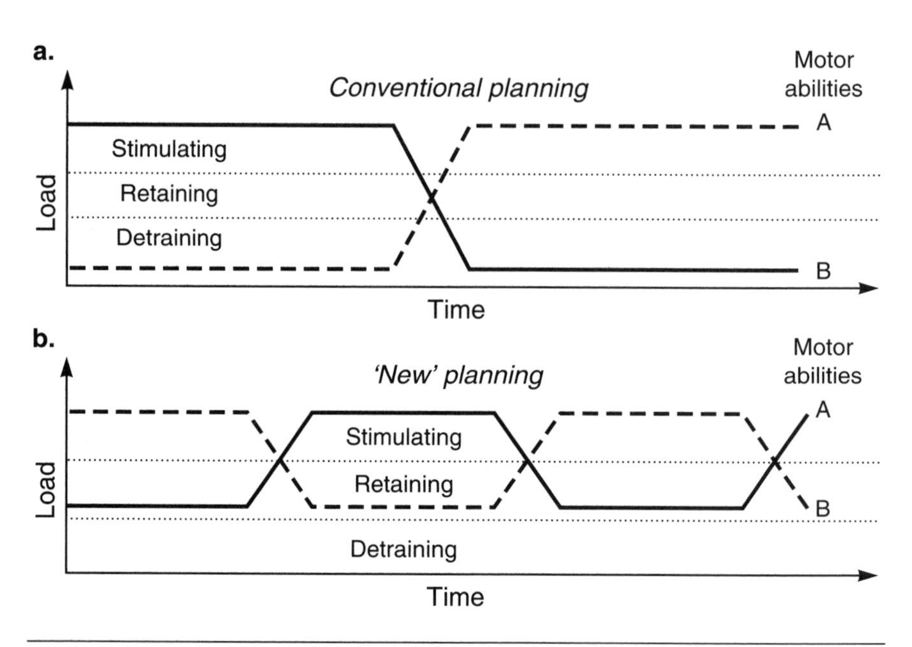

Figure 5.5 Two variations of timing training loads during a preparatory period (preseason). Two motor abilities, A and B, are the training targets. Upper figure: long intervals of accentuated (targeted) training with stimulating and detraining loads. Lower figure: short intervals of the targeted training with stimulating and retaining loads.

needed to fuse all the partial improvements into an athlete's preparedness for high-level competition.

Another training strategy has been developed in the last 15 years. The strategy is based on two ideas:

- the sequential, or even simultaneous, development of specific motor factors with frequent, intermittent changes in training targets (Figure 5.5b), and
- maintenance of the nontargeted motor abilities with retaining loads.

This approach (training various motor abilities sequentially with frequent intermittent changes of targets) is used typically with 2-week intervals, "half-mesocycles." Training targets are changed intermittently every 2 weeks. This strategy is used by athletes participating in Nordic combined competitions (cross-country skiing 15 km plus ski jumping). The skiers train in 2-week phases. During the first 2-week period, cross-country skiing is the main object of training with ski jumping loads only at the retaining level; this is followed by 2 weeks of training ski jumping (at a stimulating load) with low retaining loads in cross-country skiing, and so on.

The term "simultaneous training" means, in this case, as close in time as possible: either on the same training day, in the same microcycle, or in intermittent microcycles (but not in the same training workout). The saying is "all as close together as possible." This strategy has been successfully utilized in several power sports, for instance with soviet/Russian hammer throwers. The contributing motor abilities (maximal strength, rate of force development, power) are trained during the same microcycle with this approach.

The ideas for periodization that we have been looking at are realized in training programs in a multitude of ways.

■ *Continuous Training Is a Must*

Long breaks are customary in education. Vacations don't harm students' acquisition of knowledge or impair their intellectual abilities. After a break, they are able to study hard and learn at a faster pace.

A human body, however, behaves differently. Long breaks in training ruin physical fitness and athletic performance. De-adaptation inevitably takes place. Detraining occurs. After a prolonged period of inactivity, an athlete has to start from a decreased level of physical fitness. Time and effort is unnecessarily spent on recovering the prebreak level of fitness. If not for the break, the same efforts would be spent on increasing, not restoring, fitness. As in mountaineering, if you want to scale the summit of a high mountain, why get halfway up the mountain, go back down and then climb the whole mountain?

Prolonged interruptions in training are not good for an athlete's health. It takes time to become accustomed to regular physical exercise and also to become unaccustomed to habitual activity. Sharp decreases in an athlete's activity level offer no benefit. In fact, there is an added risk of injury, for two reasons. Various motor abilities are retained differently. Some are lost quickly and some are more stable. The new imbalance of motor abilities, for instance between high strength and decreased flexibility and relaxation, may provoke trauma. In addition, athletes are often not psychologically attuned to their new condition. They are likely to overestimate their current potential. If they try to perform as before, they may get injured.

The NCAA rules do not take full account of these natural requirements. The rules limit organized practice activities to 22 to 24 weeks a year (or 144 days) and encourage intermittent rather than continuous year-round training. Voluntary individual workouts initiated by the student-athletes and not supervised by coaches are important to maintain the previously attained fitness level. A coach is only permitted to design a general individual workout program for a student-athlete (not a specific workout program for specific days).

A better plan to educate student-athletes would be to emphasize the harm to their athletic preparedness and health of sudden changes in activity level. The athletes should be familiar with the main principles of training and should understand the personal training philosophies of their coaches.

In addition, the coach needs to design the individual workout program for each student who requires such guidance, make sure that the program is understood, and advise students about safety measures during voluntary individual workouts.

If the objective of the individual training program is limited to retaining a general fitness level, student-athletes should take several steps. They need to monitor body weight, maintaining a proper balance between overall energy expenditure and the number of calories supplied with food. The body weight must be kept constant; only a 2- to 3-kg gain is permitted. These athletes should also do calisthenics (strengthening and stretching exercises) and perform an aerobic activity to provide the minimal combined load required to retain fitness. Muscular strength, flexibility, aerobic capacities, and stable body weight must be maintained.

The laws of physical training must be obeyed if one wants to be successful in sport. The need for continuous training is one such law. If student-athletes seek to become elite athletes, possess the proper experience and knowledge, have access to practice facilities, and take safety precautions against trauma, their individual workout programs may be designed to enhance preparedness rather than only maintain it. In this case, the training should continue to follow the standard schedule with adjustments made to accommodate the athlete's responsibilities, for instance an exam schedule.

Strength Training in Macrocycles

Proper timing is vitally important for effective strength training. The timing of strength training in macrocycles, that is, in periods that are relatively long (several months), is only indirectly influenced by the exercise-rest paradigm and by the desire to avoid premature fatigue. Other facets of training become more important. In macrocycles these typically are the following:

- Demand for variability of training stimuli
- Delayed transformation of a training load (into fitness development)

- Delayed transmutation of nonspecific into specific fitness, and
- Training residuals

Variability of Training Stimuli

Demands for *variability* in macrocycles are met by changing exercise programs and training methods. Exercises themselves, not just the quantitative parameters of training routines (training load, volume, intensity), must be periodically changed to avoid accommodation. The general idea is simple. As a result of accommodation, a standard training program (same exercises, similar training load) very quickly leads to slow, or no, strength gains. To activate new steps in an adaptation, the program must be changed in one or both of two ways: increasing the training load or changing the exercise complex. There are limits to increasing the training load (because of staleness and time constraints), so changing exercises is preferable. This strategy has proven its effectiveness in the preparation of many international-caliber athletes.

The training of the best soviet/Russian hammer throwers, who have dominated world, Olympic, and European competitions over the last 30 years, is a good example of this strategy. A total of nearly 120 specific exercises were selected or invented for training and were distributed into 12 complexes with 10 exercises per complex. Each complex was used, depending on the individual peculiarities of an athlete, for 2 to 4 months and after that was replaced by another complex. The same complex was performed only once in the 2- to 4-year period. The most efficient exercises (for a given athlete) were used in the year of the most important competition (e.g., Olympic games). The athletes performed hammer throws with maximal effort almost every training day. When a strength complex was changed, performance results in hammer throwing slightly deteriorated. They began to improve, however, after a period of initial adaptation to the new load (Figure 5.6).

Strength training methods (submaximal effort, repeated effort, maximal effort) are used in different proportions within a macrocycle. Conventionally, a preparatory period begins from a mesocycle centered mainly on the submaximal effort method (the lifting of nonmaximal loads an intermediate number of times, not to failure) and the repeated effort method (maximal number of repetitions in a set). Then the athlete shifts into the maximal effort method, increasing the lifted weight and decreasing the number of repetitions per set. The strategy is to initially prepare and develop the musculoskeletal system (peripheral factors) and then improve neural coordination. This conventional paradigm has been substantially unchanged since 1980. A new tendency is to alternate or vary the training methods several times during the macrocycle. Mesocycles (4 weeks) or half-mesocycles (2 weeks), during which the methods of

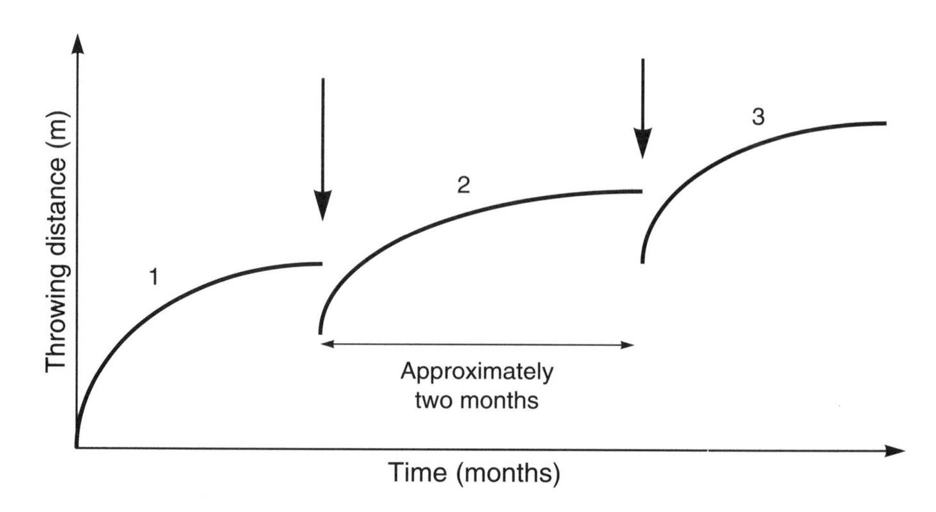

Figure 5.6 The influence of periodic changes in exercise complexes (vertical arrows) on the performance of hammer throwers. Note. The diagram is based on the concept developed by the U.S.S.R. National Olympic team head coach A.P. Bondarchuk, 1980.

repeated or maximal efforts are emphasized, follow each other in sequence.

Delayed Transformation

Because of the time delay between an increase in training load and improvement in performance, the training load should be decreased before an important competition (*the period of delayed transformation*). In essence, this is the time an organism needs for rest and adaptation.

The length of the transformation period is positively correlated with the amount of the training load, especially with the increment of the training load over the load previously employed. The higher the training load increment, the greater the time needed to adapt and the longer the period of transformation. The duration of the precompetition phase, when loads are relatively low, is typically one mesocycle or approximately 4 weeks. However, if the training load is sharply increased with the use of several stress microcycles, the precompetition phase may last up to 6 and even 7 weeks. Conversely, when the training load is mildly enhanced, the duration of the precompetition phase is around 2 weeks.

In comparison with the preparatory phase, a precompetition phase for elite weight lifters contains fewer training sessions per week (5–10 instead of 8–15), fewer exercises per workout (1–4 instead of 3–6), and fewer sets per exercise (3–5 instead of 4–8). A primary issue during this period is good rest and full restoration between workouts.

Delayed Transmutation

As the time leading up to an important competition decreases, the strength exercises should become more specific. This refers to the delayed transmutation of *nonspecific* fitness acquired in assistance exercises (relative to a main sport skill) into *specific* fitness.

Training Residuals and Retaining Loads

The level of strength an athlete has achieved can be maintained during the season (the competition period of a macrocycle) by retaining loads. Two short (30–40 min) heavy resistance workouts per week usually provide a load of sufficient magnitude. Exercising twice a week makes it possible to preserve, but not improve, the athlete's strength during the whole season.

The total training load per macrocycle is high for elite athletes and has shown a general trend toward growth (Figure 5.7 shows training loads of the Bulgarian national team). The best weight lifters of the 1960s lifted a barbell less than 10,000 times during a 1-year period:

- Yuri Vlasov, 1960 Olympic champion in the super heavy weight category—5,715 repetitions a year;
- Leonid Zhabotinsky, 1964 Olympic champion in the super heavy weight category—5,757 repetitions;
- Yan Talts, 1972 Olympic champion—8,452 lifts.

In the 1973-1976 Olympic cycle, the average number of repetitions a year for a member of the soviet national Olympic weight-lifting team was 10,600. During the 1985-1988 quadriennal cycle, it was 20,500.

For elite athletes, the training load, expressed in tons, varies substantially during year-round preparation (Figure 5.8). However, contrary to common belief, the average weight lifted (the total weight divided by the number of lifts) is rather constant. Why? Because changes in the exercises used correlate with changes in the methods of strength training. Recall that loads of 1 to 2 RM are lifted primarily in main sport exercises while a greater number of repetitions is typical for assistance exercises (see chapter 4). If an athlete during an accumulated mesocycle decreases the weight lifted in the clean and jerk and performs many barbell squats, the average weight may not change; the decrease of load in one exercise (the clean and jerk) is outweighed by the high load lifted in squats.

The rule of 60% is recommended for use in planning a macrocycle. The load in a mesocycle with minimal load is approximately 60% of a maximal mesocycle load, provided that the mesocycles are equal in duration.

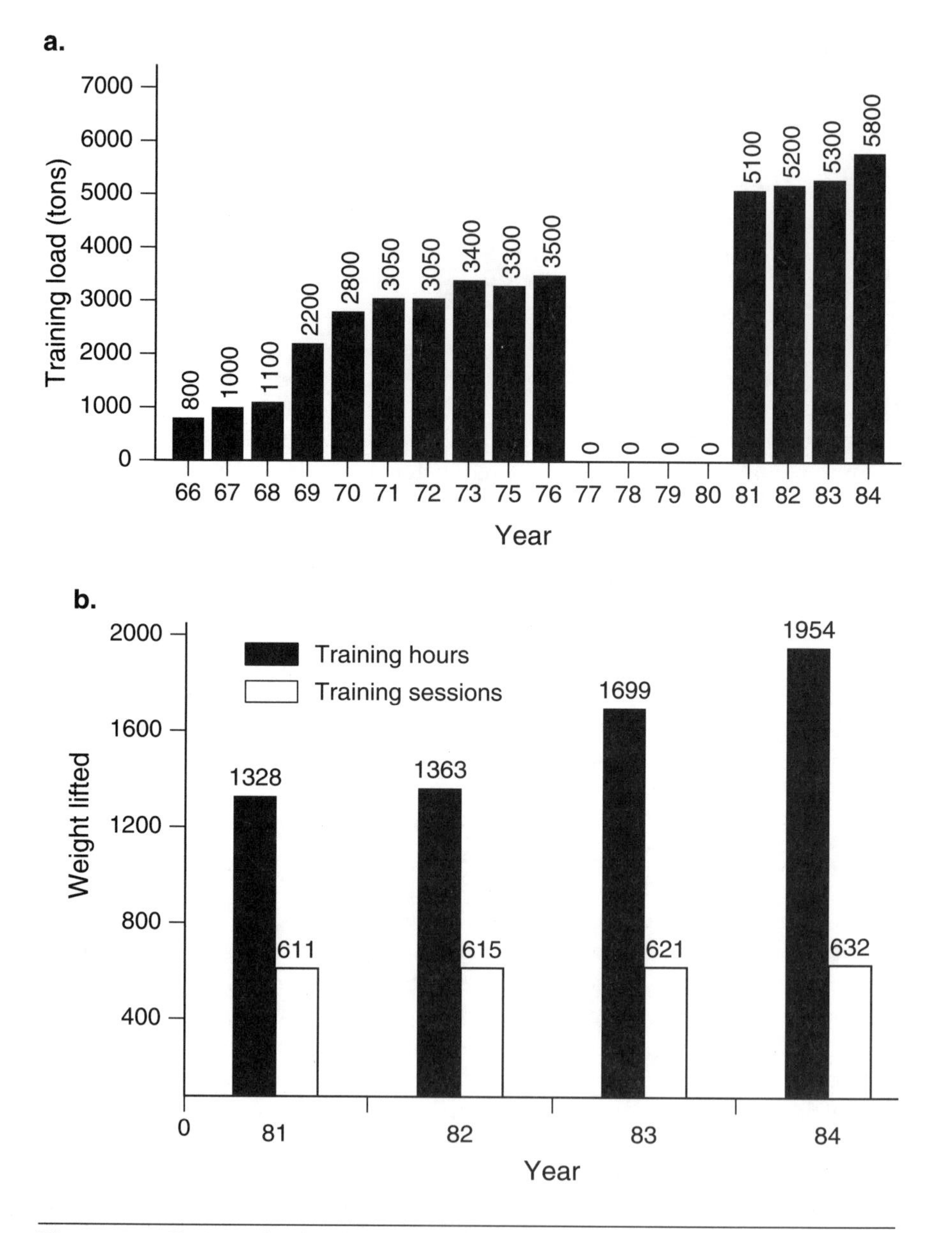

Figure 5.7 Training loads of the Bulgarian national weight-lifting team. (a) Total weight lifted; (b) number of training hours and workouts. Note. The data are from *Training of Weight Lifters* by I. Abadjiev and B. Faradjiev, 1986, Sofia, Bulgaria: Medicina i Fizkultura.

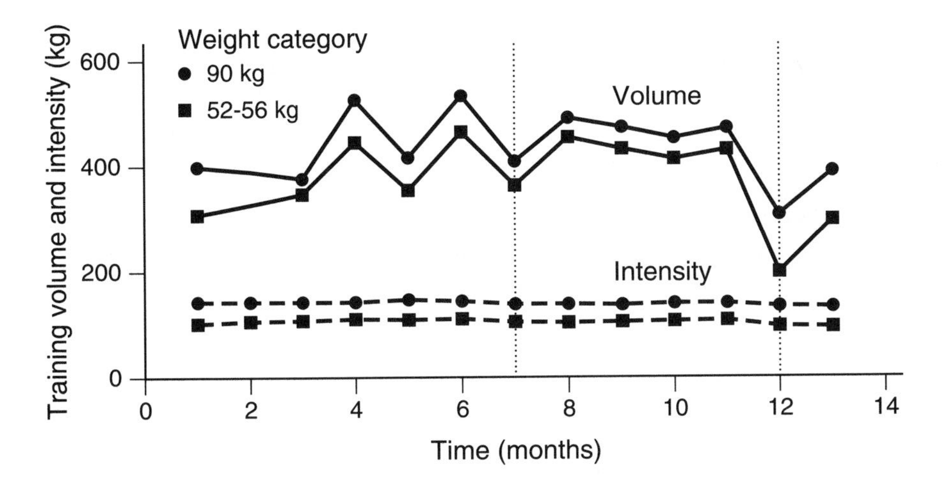

Figure 5.8 The distribution of training volume (tons, two top curves) and average weight lifted (kg, two bottom curves) by the U.S.S.R. National Olympic team during year-round training. Average data of athletes in weight categories 52–56 kg ($n = 4$) and 90 kg ($n = 3$). Vertical dotted lines show the times of important competitions. Note. Data are from *Preparation of the National Olympic Team in Weight Lifting*, 1984, annual report #85-012, Moscow: All-Union Research Institute of Physical Culture.

Summary

The timing of training includes the spacing of work and rest intervals as well as the sequencing of exercise. Training can be divided into structural units consisting of the (a) training session (workout), (b) training day, (c) microcycle, (d) mesocycle, (e) macrocycle, (f) Olympic cycle (quadrennial cycle), and (g) long-lasting, or multiyear, training.

Short-term planning refers to the planning of workouts, training days, microcycles, and mesocycles (typically 2–6 weeks). A general principle of short-term training design is that the effects of fatigue from different types of muscular work are specific. Thus an athlete who is too tired to repeat one exercise in an acceptable manner may still be able to perform another exercise to satisfaction. Training too many motor abilities during the same workout, microcycle, or mesocycle lessens effectiveness. Two or three main targets are plenty. Try to balance the number of training targets in these cycles to enhance performance growth while also planning the schedule to avoid the superimposition of fatigue traces from individual workouts and the hazard of staleness.

The general idea in planning workouts and training days is to have athletes do as much work as possible while they are as fresh as possible. Unlike the situation with endurance training, it isn't necessary that they become exhausted in a heavy resistance workout. To prevent early fatigue, rest intervals between sets, especially when trainees are working with heavy weights, should be long (about 4-5 min). During training days, distributing the training volume into smaller units has a definite advantage, provided that the time intervals between workouts is sufficient for restoration.

In properly sequenced exercise, the athlete performs the most valuable exercises—those requiring fine motor coordination and maximal neuronal output—in a rested state. To prevent premature fatigue, include main sport exercise before assistance exercises; use dynamic, power-type drills before slow exercises (such as squats); and exercise larger muscle groups before smaller ones.

The method of maximum efforts is recognized as the most efficient training method and should be practiced at the beginning of a training workout, following warm-up. Pyramid training is ineffective and even detrimental. Mixed training sessions that include a strength routine section are less effective for strength development than special heavy resistance workouts. The same holds true for circuit training.

In planning heavy resistance protocols in micro- and mesocycles it is important to assign adequate rest between exercise periods and to balance the stability of a training stimulus (to call forth an adaptation) and its variation (to avoid premature accommodation and staleness).

Adequate recovery during a microcycle is achieved by rest-exercise alternation and proper exercise sequencing. To retain the attained strength gains, schedule at least two training sessions per week. The variability of training programs during micro- and mesocycles is realized through changes in training load (not exercise complexes). One stable complex of exercises should be performed through a mesocycle (to elicit an adaptation). The empirical rule of 60% has stood the test of time: The training volume of a day (microcycle) with minimal loading should be about 60% of the volume of a maximal day (microcycle) load.

Medium-term planning (periodization) deals with macrocycles. When you periodize, you divide the training season, typically 1 year long, into smaller and more manageable intervals (periods of training, mesocycles, and microcycles), with the goal of getting the best performance results during the primary competition(s) of the season.

In periodizing, allow for delayed transformation. During periods of strenuous training, athletes cannot achieve the best performance results. They need an interval of relatively easy exercise to realize the effect of previous difficult training sessions. The adaptation occurs (or is manifested) during unloading, rather than loading, periods. Another phenomenon that you need to consider is delayed transmutation. A special training routine is needed to transmute the acquired motor potential into athletic performance. This goal is realized during highly specific training in transformation mesocycles. Finally, it is important to take training residuals into account when you plan for the medium term. De-adaptation, as well as adaptation, takes time. The time course of training should be based on the duration of the immediately preceding period of training, the training experience of the athletes, the motor abilities being targeted, and the training volumes during training mesocycles.

A good periodization plan is a subtle trade-off among conflicting demands. The conventional approach has been to solve the problem sequentially, for instance to begin off-season preparation with nonspecific strength training and after that change to a highly specific technique routine. A more recent strategy is to sequentially develop specific motor abilities with frequent intermittent changes in training targets and to maintain the nontargeted motor abilities with retaining loads.

The timing of strength training in macrocycles is only indirectly influenced by the exercise-rest paradigm and by the desire to avoid premature fatigue. Other training facets also influence timing, including the demand for variability of training stimuli, delayed transformation of training load (into fitness development), delayed transmutation of a nonspecific fitness acquired in assistance exercises (relative to a main sport skill) to specific fitness, and training residuals.

CHAPTER 6

Strength Exercises

The first problem a coach encounters in planning a training program is exercise selection. The alternatives seem innumerable: free weights, exercise machines, isometrics, uphill ambulation with an additional load, dropping jumps, self-resistance exercises, and so on. In this chapter you will read about various classes of exercises used for strength enhancement.

Classification

Strength exercises typically are classified according to the change in muscle length. They may be *static*, or *isometric*, which literally means constant (*iso*) length (*metric*), or *dynamic*, a category further divided into exercises with *concentric, eccentric,* or *reversible* muscle action. Dynamic exercises are also sometimes labeled *isotonic* (from *iso,* meaning constant, and *tonic,* meaning tension). The underlying assumption is that the muscle produces an unvarying amount of tension while shortening as it overcomes a constant resistance. This is not the case for intact muscles. If external resistance (weight lifted) is constant, the tension exerted by a muscle varies during shortening because of factors such as the change in muscle moment arm.

Among dynamic exercises, one special class is termed *isokinetic* (*iso* is for constant and *kinetic* is for speed). During isokinetic action, the speed of movement is constant, regardless of muscle tension. (The term *isokinetic,* unfortunately, is not strictly defined. *Speed of movement* may refer either to rate of change in muscle length, velocity of the load being lifted, or angular velocity of the joint.) Special equipment, usually expensive, is necessary for proper isokinetic training.

Because dynamic exercises with concentric muscle action are much more popular in athletic training than other types of exercise, these will be our focal point.

Strength exercises can also be grouped according to the muscles involved in the action (e.g., abdominal exercises, leg exercises). The strength of different muscle groups often varies greatly in one person. An athlete can have high strength in one movement, for instance in the leg extension, but be relatively weak in another, such as pull-ups. The comparative strength of different muscle groups is called *strength topography.*

In addition, strength exercises are often classified according to their specificity as (a) *nonspecific* (e.g., barbell squats for javelin throwers or baseball pitchers); (b) *specific* (exercise for muscles specifically involved in a throwing task, e.g., an exercise for shoulder extensors in which a weight is pulled to imitate an overhand throwing motion); and (c) *sport exercise with added resistance* (e.g., overhand throwing of heavy objects).

Exercise Selection for Beginning Athletes

With beginning athletes, especially youngsters, *strength topography* is the main focus in the proper selection of strength training exercises. The most important muscle groups should be chosen and trained. The following recommendations are offered as a rule of thumb:

1. Strengthen the muscle group that, if weak, can increase the risk of trauma (for instance, neck muscles in wrestling and tackle football).
2. Train large, proximally located muscles, especially in the trunk area, with the abdominal wall muscles and spine erectors as a primary choice.

3. Increase strength in sport-related movements to a level that permits sport technique acquisition without technical mistakes.
4. Have athletes perform movements through the entire range of angular joint motion. The submaximal effort and repeated effort methods only, not singular maximal efforts, should be employed.
5. The so-called "three-year rule" is popular among experienced coaches. According to this rule, an athlete should use strength-specific exercises and exercises with a barbell, such as barbell squats, only after 3 years of preliminary general preparation.

■ *Which Muscle Groups Are Most Important? How Do We Evaluate General Strength Development?*

For more than a century, hand grip strength was commonly used to estimate the strength development level of various subjects and populations. Is grip strength a valid test for whole body strength? In the grip test, the thumb produces force against the force generated by the other four fingers. Since the four fingers together can generate greater force values than the thumb alone, in reality the strength of only the thumb is measured in the test. Is the strength of the thumb so important in athletics and everyday life that it should be considered a valid, or even a unique measure of strength development? Certainly not. Which muscle groups are, then, the most important?

This question has been addressed in several investigations. Ideally, a small number of muscle groups or exercises could be found that would represent with maximal precision the achievements in a large test battery. To find such a set of muscles, groups of subjects were given many (up to 100) strength tests. Statistical analyses were employed to find the most representative (important, valid) muscle groups and tests. The results led to recommendations for choosing the most representative muscle groups given a limited number of tests.

If you are selecting up to five muscle groups, use the abdominal muscles, spine erectors, leg extensors, arm extensors, and pectoralis major. When you are limited to two tests, use measures of strength lift on a high bar with overturn and forcible leg extension (e.g., squatting on one leg).

Exercise Selection for Qualified Athletes

Selecting strength exercises for qualified athletes is substantially more complex than for beginners. The general idea is simple: Strength exercises must be *specific*. This means that training drills must be relevant to the demands of the event an athlete is training for. Strength training drills must mimic the movement pattern that the pertinent sport skill actually entails.

However, the practical realization of this general idea is not easy. Coaches and athletes have made many efforts to find the most effective strength training drills for various sports. The main requirements of this task are described below. We will consider yielding strength and strength in reversible muscle action as separate motor abilities and discuss them later.

Working Muscles

The requirement regarding working muscles is most evident and simple. The same muscle groups must be involved in both the main sport event and in the training drill. For instance, heavy resistance exercises for the improvement of paddling in canoeing should focus on the muscles utilized in the motion patterns associated with the paddle stroke.

Unfortunately, this obvious requirement very often is not satisfied in athletic practice. Coaches and athletes often employ exercises and training equipment that are not specific—that do not involve the muscle groups active in the main sport movement. For instance, in swimming, the athlete's hand moves along a complex curvilinear trajectory that includes inward and outward motions (see Figure 6.1). The resistance vector occurs in a three-dimensional space (Figure 6.1a). During dryland training, however, swimmers typically use exercise devices with linear, straight-back pulls (Figure 6.1b). Muscle activity patterns during such training are distinctly different from those experienced while swimming. It is preferable to mimic the three-dimensional hand resistance that occurs in swimming by using two- and three-dimensional exercise devices (Figure 6.1c).

Muscle activity in the same exercise can vary if the performance technique, such as the body posture, is changed. This is illustrated in Figure 6.2. An athlete performs shoulder squats with a barbell using different lifting techniques. Not only does the level of muscle activity change, but also the involvement of specific muscle groups; the knee extensors are used in some instances and the knee flexors in others.

Four techniques are employed to identify the working muscle groups:

- Muscle palpation. Muscles that become tense are the "involved" muscles and these should be trained with heavy resistance.
- Intentional inducing of *delayed muscle soreness* (i.e., the pain and soreness that occur 24 to 48 hr after training workouts). To this end, a coach intentionally overdoses the training load during the first workout with new drills. The painful muscles are then identified as the working muscles.

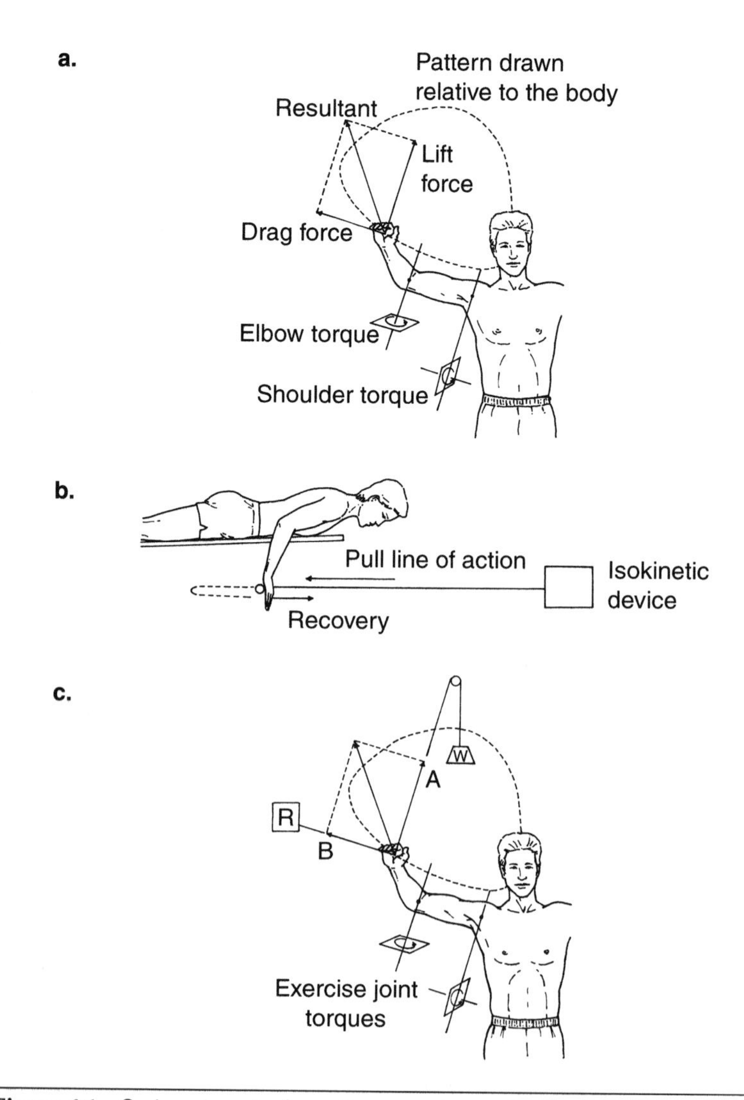

Figure 6.1 Swimming stroke patterns and exercise devices for dryland training. (a) Arm trajectory and propulsive forces in the breaststroke. (b) An exercise device with a straight-line pull. The device provides only single-line, one-dimensional resistance. (c) Two-dimensional resistance force. Two resistances, A and B, are provided that duplicate the lift and drag components of propulsive swimming force. Note. From "Specificity of Strength Training in Swimming: A Biomechanical Viewpoint" by R.E. Schleihauf, in A.P. Hollander, P. Huijing, and G. de Groot (Eds.), *Biomechanics and Medicine in Swimming* (pp. 184-191), 1983, Champaign, IL: Human Kinetics. Copyright 1983 by Human Kinetics. Reprinted by permission.

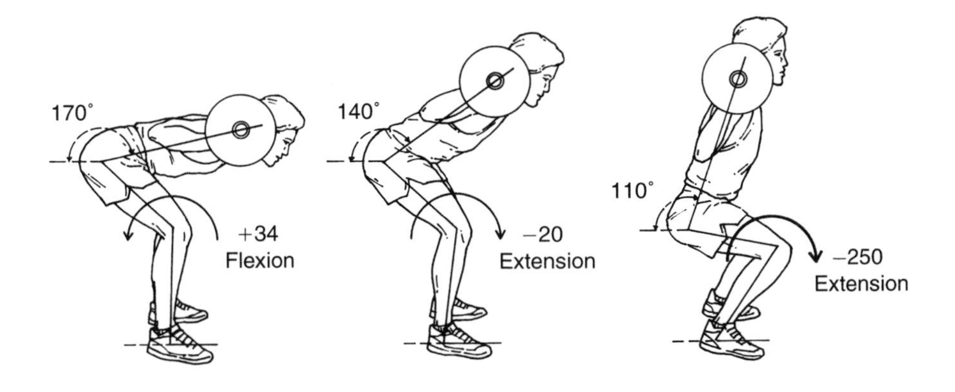

Figure 6.2 Net muscle moments in the knee joints (Nm) during squatting with an 80-kg barbell. Both the magnitude and direction (flexion or extension) of the moment are altered when the athlete's posture is changed. Note. The data are from *Force-Posture Relationships in Athletic Movements* by V.M. Zatsiorsky and L.M. Raitsin, 1973, technical report, Moscow: Central Institute of Physical Culture.

- Biomechanical analysis of the joint torques similar to that presented in Figure 6.2. The method is good but too complex for practical use.
- Registering muscle electric activity (EMG). This method is superior, but special equipment and technical personnel are needed for this type of analysis.

■ *Are Chin-Ups and Dips Equally Effective for Various Sports?*

A conditioning coach is working with several varsity teams: football players (running backs and wide receivers), volleyball players, swimmers, and rowers. She has been asked by the head coaches to pay special attention to the enhancement of arm strength. Her general philosophy is to use strength exercises that are specific to relevant sports. In addition, the time available for strength training is limited. Thus, she must recommend only the most efficient exercises. The following variants of chinning (on a horizontal bar) versus dipping (on a parallel bar) distribution are considered (%): (a) 100/0, (b) 70/30, (c) 50/50, (d) 30/70, and (e) 0/100.

What is your choice? Please substantiate it.

Other requirements for exercise selection, described later in this chapter, are less obvious. These are based mostly on factors determining the amount of muscular strength available in various motor tasks (see chapter 2).

Type of Resistance

In terrestrial movements, the weight or mass of an object (an implement, barbell, one's own body)—or a combination of weight and mass—usually serves as resistance. In aquatic sports, such as swimming, rowing, kayaking, and canoeing, the resistance is determined by hydrodynamic principles. If the training drill resistance is different from the resistance in the sport event the athlete is training for, both the force production (recall Figure 2.3) and the pattern of muscle activity are also different.

In *terrestrial movements,* when an object of given mass (a throwing implement or an athlete's extremity) is accelerated, the burst of muscle action is "concentrated" both in time and in space. Thus, the muscle action is of short duration and the maximal force is developed in a specific body position. If other types of external resistance are used in training, such as devices with hydraulic resistance, rubber cords, and isokinetic machines, the maximal force is developed either throughout the whole range of the angular motion or in a body position different from the position used in the sport event. The muscle action is not concentrated in time but instead tends toward protracted activation. Such exercises are not specific for on-land athletic events. The first choice for these sports is exercises that use free weights, the body mass, or both as resistance.

In *aquatic sports,* water resistance (while pulling through the water) increases with velocity (an example of *mechanical feedback;* see chapter 2). The relationship is quadratic, meaning that external force applied by the athlete is proportional to the squared velocity of the arm or paddle relative to the water. Athletes contract their muscles in a protracted way. This type of activity must be simulated during dryland training. The best choice here is training equipment in which the resistance is proportional to the velocity squared. This equipment is rather expensive and impractical, though.

There are also machines in which the resistance (F) is either proportional to the velocity of movement or constant through the whole range of the motion. In the first type of device, oil viscosity is used as a resistant force. In an apparatus based on hydraulics, the oil is squeezed from one chamber to another through an adjustable orifice. The greater the velocity of the forced oil displacement, the greater the resistance offered by the training device. In devices of the other type, dry (Coulomb's) friction serves as resistance. The force (F) is constant if the velocity (V) is not equal to zero (F = Constant, if V > 0). The force may be changed from zero to F at zero velocity. These two types of devices should be used as a second choice.

Time (and Rate) of Force Development

Because of the explosive strength deficit (ESD; see chapter 2), maximal force F_{mm} cannot be attained in the *time deficit zone*. If the training objective is to increase maximal force production (F_{mm}), there is no reason to use exercises in the time deficit zone, where F_{mm} cannot be developed. Furthermore, heavy resistance exercises are not very useful for enhancing the rate of force development in qualified athletes (Figure 6.3).

If the general objective of training is to increase force production in explosive types of movement, in principle this can be done in one of two ways. One option is to increase maximal force F_{mm}. This strategy, however, brings good results only when the ESD is substantially less than 50%. As an example, imagine two athletes who put a shot with a force of 500 N. The first athlete can bench press a 120-kg barbell (roughly 600 N per arm). The ESD for this athlete is $[(600 - 500)/600] \cdot 100 = 16.6\%$. This is an extremely low value for shot-putting. The athlete has a great potential to improve performance by increasing F_{mm}. Lifting a 200-kg barbell in the bench press will surely lead to improvement in this individual's performance. For the other athlete, 1 RM in the bench press is 250 kg. The ESD is $[(1250 - 500)/1250] \cdot 100 = 60\%$. Further improvement of this athlete's maximal bench press, say to 300 kg, will not result in improvement in shot-putting performance.

The second option for training to enhance force production is to increase the rate of force development. Heavy resistance exercises are not the best choice in this instance, especially for elite athletes. Special exercises and training methods are a better alternative.

■ *Are Strength Exercises Equally Useful for All Athletes?*

Two athletes of similar body dimensions possess equal achievements in the standing vertical jump. Their performances in barbell squats, however, are different. Athlete P squats a barbell equaling his body weight (BW). Athlete Q can squat a 1.5-BW barbell.

For which of these athletes will barbell squatting be more beneficial? Why?

Velocity of Movement

The effects of a strength exercise depend on movement velocity. If exercises are performed in the high-force, low-velocity range of the force-velocity curve (Figure 6.4a), the maximal force F_m increases mainly in the

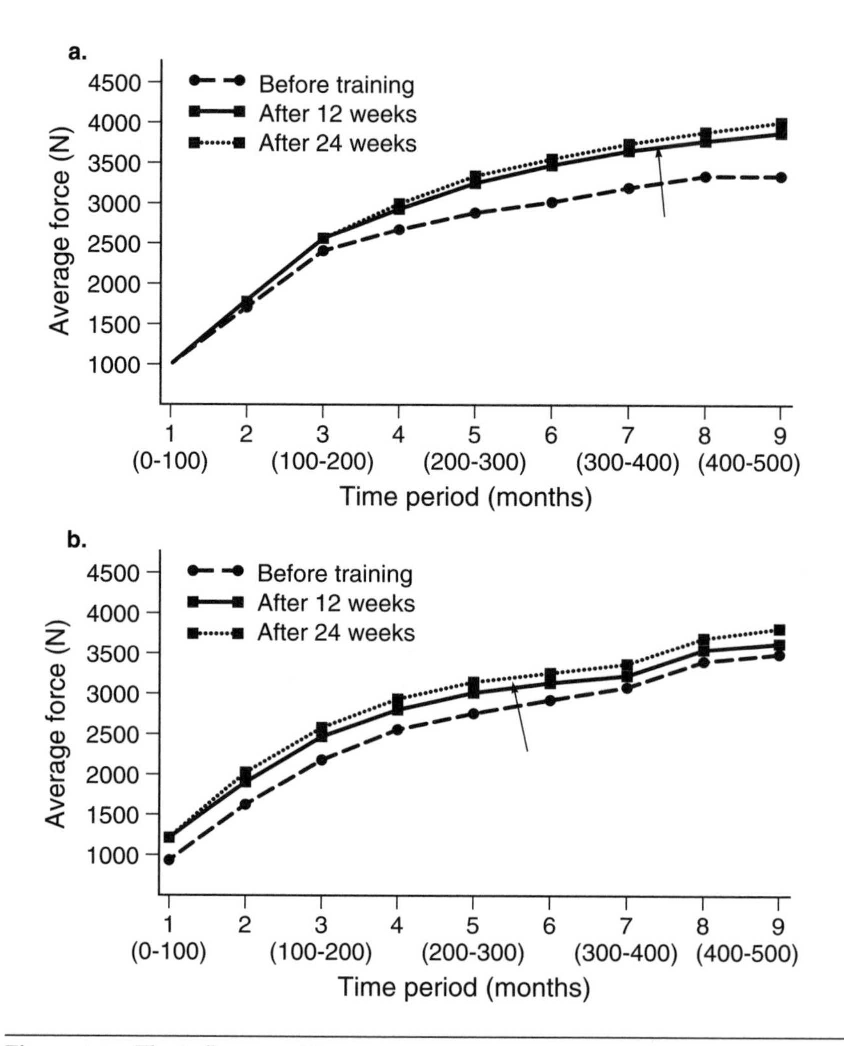

Figure 6.3 The influence of heavy resistance training (top) and dynamic (explosive, power) resistance training (bottom) on maximum strength and the rate of force development during an explosive maximal bilateral leg extension. As a result of heavy resistance training, only F_{mm}, not the initial part of the force-time curve, is enhanced. The rate of force development, especially the S-gradient, is unchanged. Note. Adapted from "Changes in Electrical and Mechanical Behavior of Leg Extensor Muscles During Heavy Resistance Strength Training" by K. Håkkinen and P.V. Komi, 1985, *Scandinavian Journal of Sport Sciences*, **7**, pp. 55-64; and "Effect of Explosive Type Strength Training on Electromyographic and Force Production Characteristics of Leg Extensor Muscles During Concentric and Various Stretch-Shortening Cycle Exercise" by K. Håkkinen and P.V. Komi, 1985, *Scandinavian Journal of Sport Sciences*, **7**, pp. 65-75. Reprinted by permission from the authors.

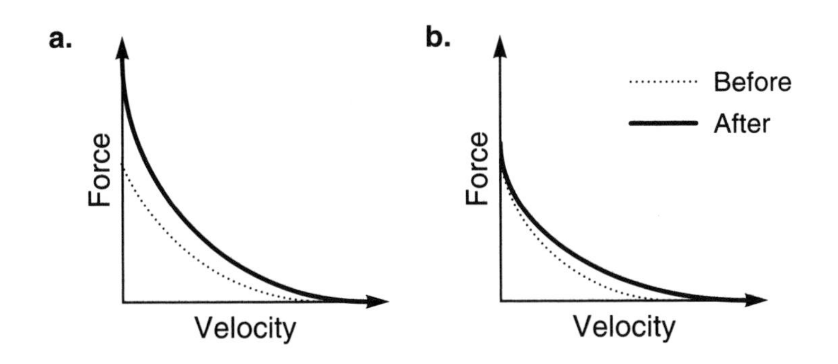

Figure 6.4 Force-velocity relationships before and after muscle power training with different loads. (a) High resistance (around 100% F_{mm}); (b) low resistance (around 0% F_{mm}).

trained range. On the other hand, if exercise is done in the low-force, high-velocity range, performance improves primarily in this area (Figure 6.4b).

These findings serve as a basis for the recommendation to develop force at speeds that approximate the athletic motion. This is particularly relevant for relatively slow movements such as those used during dry-land training for aquatic sports and endurance sports. However, if an exercise is performed in the low-force, high-velocity range, the time available for movement may be too short to develop the maximal force F_{mm}. The situation we looked at earlier, with training either F_{mm} or rate of force development, occurs (see the section on time of force development, chapter 2). The practical solution is based on these recommendations:

- Lift all weights with the maximum attainable velocity;
- For F_{mm} training, do not decrease the resistance too much; and
- For dynamic strength training, choose the amount of resistance that will produce a movement in the same velocity range as the relevant sport event.

■ *What Is the Optimal Weight of Medicine Balls?*

Medicine balls are popular implements for training throwing tasks. In one study, the optimal weight of medicine balls for dryland training of water polo players was determined. The ball velocity in natural conditions (in the water) served as a criterion. The results showed that the optimal medicine ball weight for dryland training is approximately 2.0 kg. Both the correlation with throwing velocity in water conditions and the gain in velocity due to training were highest with medicine balls of this weight.

Don't misinterpret the suggestion to perform strength exercises with maximum velocity as advice to execute these exercises with *high frequency* (high number of repetitions per minute). Too high a frequency has been shown in several experiments to impede strength gain. If movement frequency is in the medium range, its precise value is of no importance. In one experiment, for instance, strength gain in the bench press was similar when a barbell was lifted 5, 10, or 15 times a minute but was much smaller for athletes who performed the lifts with the maximum possible frequency.

Force-Posture Relationship

As described in chapter 2, certain relationships exist between body posture (i.e., joint angle) and the maximal strength that can be developed at this body position (these relationships are called *human strength curves*). Maximal muscular strength varies over the full range of joint motion. The interplay of changes in muscle lever arms and muscle force production determines this effect.

The magnitude of weight that an athlete can lift in a given motion is limited by the strength attainable at the *weakest point* of the full range of joint motion. In other words, the weakest point of a muscle group determines the heaviest constant weight that can be lifted. If the constant external resistance (such as a barbell of a given weight) is used in heavy resistance training, the muscles are maximally activated at only the weakest point of the motion. For instance, there is a threefold difference in the maximal force that can be developed at different angles of hip flexion (Figure 6.5). If someone lifts the maximal weight that is equal to 100% of F_m at the weakest point of movement (at a 70° hip-joint angle), the hip flexor muscles are taxed to only 33% of maximal strength at the strongest point (at a 150° angle). The muscles are not required to exert maximal force in this region.

Three approaches are used in contemporary strength training to manage the force-angle paradigm (the fourth "solution" is to not pay attention to this issue at all). They are the *peak-contraction principle, accommodating resistance*, and *accentuation*.

The Peak-Contraction Principle

Historically, the peak-contraction approach is the oldest one. The idea is to focus efforts on increasing muscle strength primarily at the *weakest*

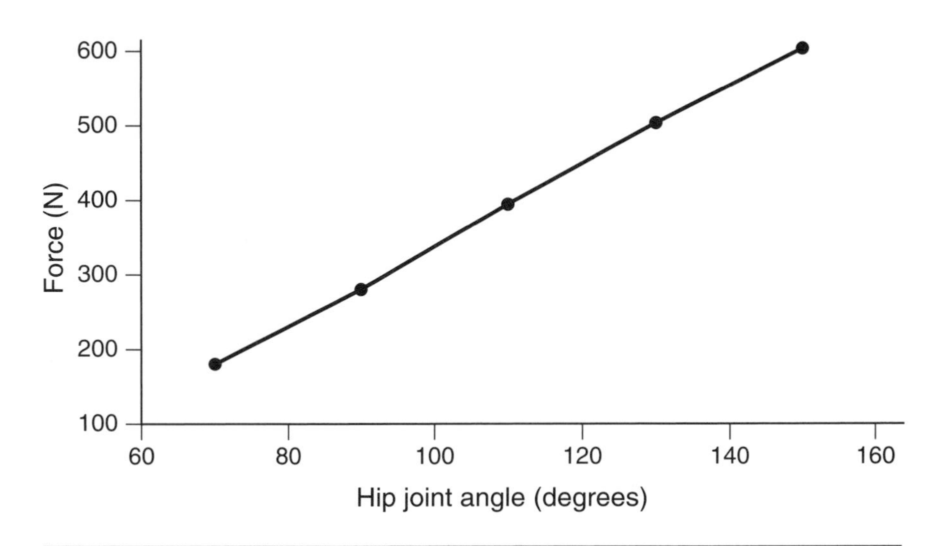

Figure 6.5 Strength curves in hip flexion. Isometric force, men. Angle of 180° is in anatomical position. Note. Adapted from "Strength Variations Through the Range of Joint Motion" by M. Williams and L. Stutzmann, 1959, *Physical Therapy*, **39**, pp. 145-152. Reprinted by permission from the American Physical Therapy Association.

points of the human strength curve. Thus the entire performance, for instance 1 RM, is enhanced. In practice, the peak-contraction principle is realized in one of three ways:

1. *Selection of a proper body position.* The resistance offered by the lifted load is not, in reality, constant over a full range of joint motion. The resistance is determined by the moment of gravitational force (i.e., by the product of weight and horizontal distance to the axis of rotation) rather than by the weight of the implement or the body part itself. The moment of gravitational force is maximal when the center of gravity of the lifted load is on the same horizontal line as the axis of rotation. In this case, the lever arm of the gravitational force is greatest. By varying body posture, it is possible to an extent to superimpose the human strength curve on the resistance curve in a desirable manner.

The peak-contraction principle is realized, if "worst comes to worst," when the external resistance (moment of gravity force) is maximal at the point where muscular strength is minimal. The corresponding body position is called the *minimax* position. The term *minimax* literally means minimum among maximums. At each of the joint angles, the strength that is maximal for this position, F_m, is developed. The minimum F_m from this set is the minimax value.

To visualize this, compare an exercise such as leg raising from two starting body positions: lying supine and hanging on a horizontal bar

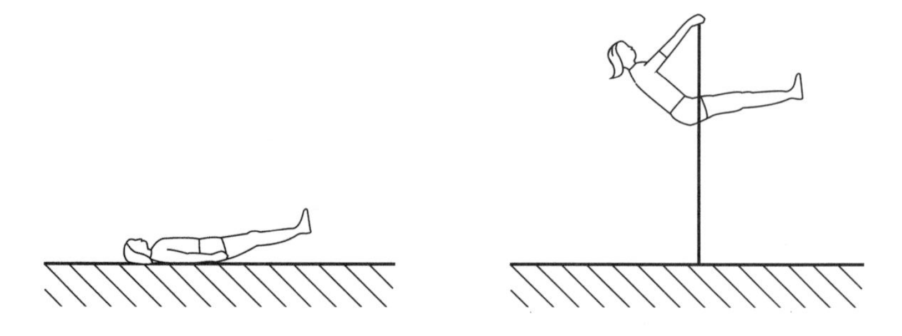

Figure 6.6 Leg raising from two different starting positions. The load is higher in leg raising performed on a horizontal bar than in leg raising from a supine position.

(Figure 6.6). The second exercise imposes a much greater demand than the first.

The resistance (moment of gravity force) is nearly equal in the two exercises and reaches its maximum when the legs are placed horizontally. However, when the legs are raised in the recumbent position, the maximal resistance coincides with the strongest points of the force-angle curve (the hip flexor muscles are not shortened). When the same leg raising is performed on the horizontal bar, the hip flexor muscles are shortened at the instant the legs cross the horizontal line. Thus, the position of maximal resistance coincides with the minimal (weakest) point on the force-angle curve ("worst comes to worst").

2. *Use of special training devices.* An example of a special device is shown in Figure 6.7. If a barbell were used in the arm curl, the maximal resistance would be at the horizontal position of the forearms. In contrast to the situation with the peak-contraction principle, the strength of forearm flexion at the elbow joint is maximal, not minimal, at this position. With the device shown in Figure 6.7, maximal resistance coincides with the weakest point on the human strength curve.

3. *The slow beginning motion.* A slow start can be used in strength drills such as the inverse curl shown earlier (Figure 4.6). The maximal resistance in this exercise is offered while the trunk is in the horizontal position. If the movement begins too fast, the lift in the intermediate range of motion is performed at the expense of the kinetic energy acquired in the first part of the movement. The erector spinae muscles, then, are not fully activated. Experienced athletes and coaches advocate a slow start for this drill.

Studies of maximal external resistance (moment of force) exerted against different points along the human strength curve (strong points or weak points) have shown that when the peak-contraction principle is employed, strength gains are higher. Thus, this training protocol has a

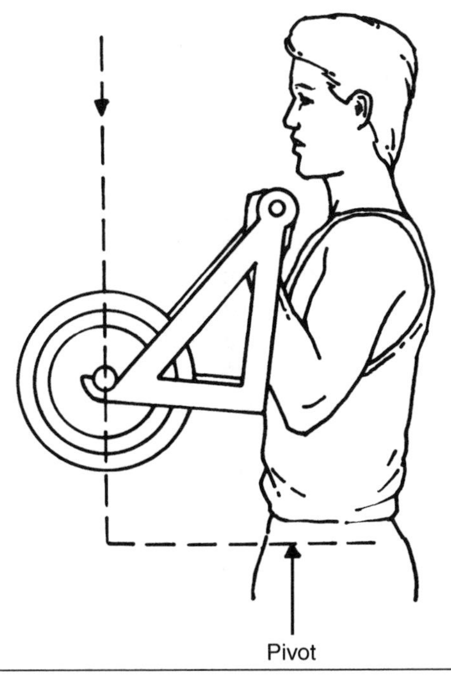

Figure 6.7 A device used to implement the peak-contraction principle. The device is employed to perform the arm curl. With this device, the highest resistance is provided at the end of the movement. When the elbow is maximally flexed, the athlete's force potential is minimal (see Figure 2.22) and the resistance is the greatest ("worst comes to worst").

definite advantage. Another merit is the relatively small amount of mechanical work performed (the total weight lifted). A disadvantage, however, is that the transfer of training to other body positions is relatively low (see Figure 1.3). A coach should consider the pros and cons of this principle before implementing it.

Accommodating Resistance

The main idea of accommodating resistance is to develop maximal tension throughout the *complete range of motion* rather than at a particular (e.g., weakest) point. This can be achieved in two ways. One type of system offers *high resistance without mechanical feedback*. In this case, the speed of motion is constant no matter how much force is developed. This principle is realized in isokinetic equipment. The movement speed on such devices can be preset and maintained (kept constant) during a motion regardless of the amount of force applied against the machine. The working muscles are maximally loaded throughout the complete range of motion. (Isometric exercises at different joint angles, in which velocity is zero, can be considered an extreme example of this approach.) Because the velocity of muscle shortening is predetermined, the training of different types of muscle fibers (fast or slow) can potentially be stressed within the framework of the isokinetic protocol.

Isokinetic training, while very popular in physical therapy, is rarely used by elite athletes. It has shortcomings besides the high cost of the equipment, which may be prohibitive. The angular velocity of movement is typically relatively low—below 360°/s (it may be above 5,000°/s in athletic movements). Most training devices are designed to exclusively perform one-joint movements that are only used sporadically in athletic training.

Another type of system provides *variable resistance* that is accommodated to either the human strength curve or movement speed. In some machines, resistance is applied in concert with the human strength curve (Nautilus-type equipment). Because of the special odd-shaped cams on these machines, the lever arm of the resistance force or applied force is variable so that the load varies accordingly (Figure 6.8).

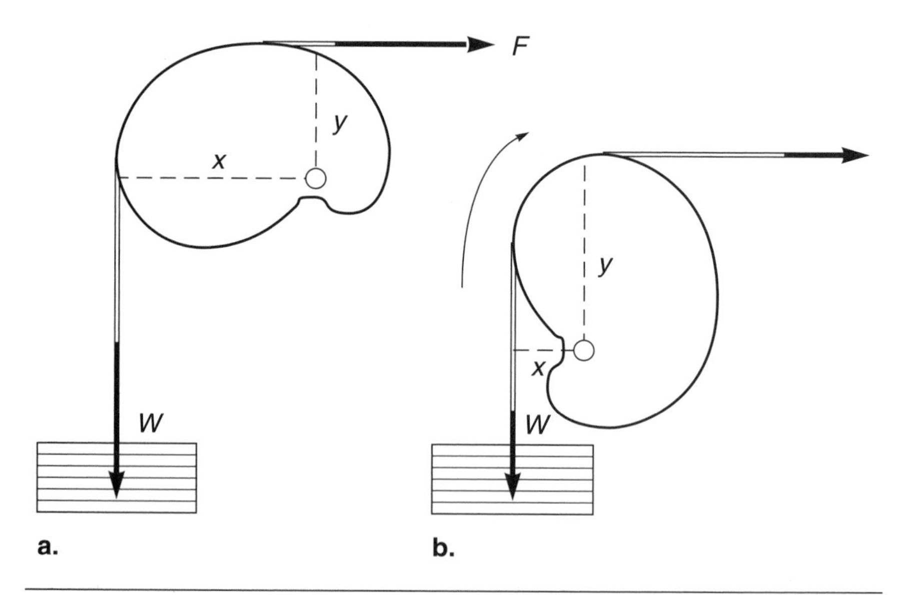

Figure 6.8 A cam with variable lever arms. In this arrangement, the moment arms of both applied force (F) and weight force (W) are variable. (a) The moment arm of the applied force (y) is smaller than the lever arm of the weight force (x), y < x; this ratio is used at points of the strength curve where a larger F can be exerted; (b) y > x; this ratio is used at the weakest points of the strength curve. Note. Adapted from *Weight Training: A Scientific Approach* (p. 84) by M.H. Stone and H.S. O'Bryant, 1987, Minneapolis: Bellwether Press.

The resistance (moment of force) varies in accordance with an athlete's capabilities. This variation, it is claimed, provides greater resistance at the joint configurations where the athlete is stronger and lower resistance at weaker positions. The user must exert maximum effort throughout the range of movement. However, the variable resistance offered by these machines is based on average data and does not match well with individual strength curves. In addition, the cams of many machines are incorrectly designed and the offered resistance, contrary to claims, does not match average strength curves.

Another type of exercise apparatus accommodates resistance to movement velocity. The higher the velocity, the greater the resistance offered by the system. These devices are typically based on hydraulic principles. The velocity of movements with hydraulic machines, in contrast to isokinetic devices, may vary depending on the strength of the trainee.

Scientific experts have often questioned the validity of claims for high exercise efficiency with accommodating resistance. Exercises performed with strength training machines are biomechanically different from natural movements and traditional exercises. Most notably, the number of degrees of freedom (permissible movement directions) is limited from six in natural movements to only one with exercise machines; the typical acceleration-deceleration pattern is also different. Though isokinetic training may have certain advantages in clinical rehabilitation settings, studies have repeatedly failed to demonstrate that accommodating resistance exercises (e.g., isotonic, variable cams) hold an advantage over free-weight exercises for increasing muscular strength and inducing muscle hypertrophy.

Accentuation

In *accentuation* the main idea is to train strength only in the range of the main sport movement where the demand for high force production is maximal. In natural movements, at least on land, muscles are active over a relatively narrow range of motion. Usually, maximal muscle activity occurs near the extreme points of angular motion. The movement of body parts is first decelerated and then accelerated by virtue of muscular forces. For instance, during the swing movement of a leg (e.g., in jumping and running), the previously stopped thigh is accelerated prior to the vertical position and decelerated afterwards (Figure 6.9a).

If the training objective is to increase the dynamic strength of the hip flexor muscles to improve velocity of the swing movement, there is no reason to increase the strength of these muscles in a range beyond the range this activity requires. An exercise that satisfies the requirement for specificity of the range of force application is shown in Figure 6.9b.

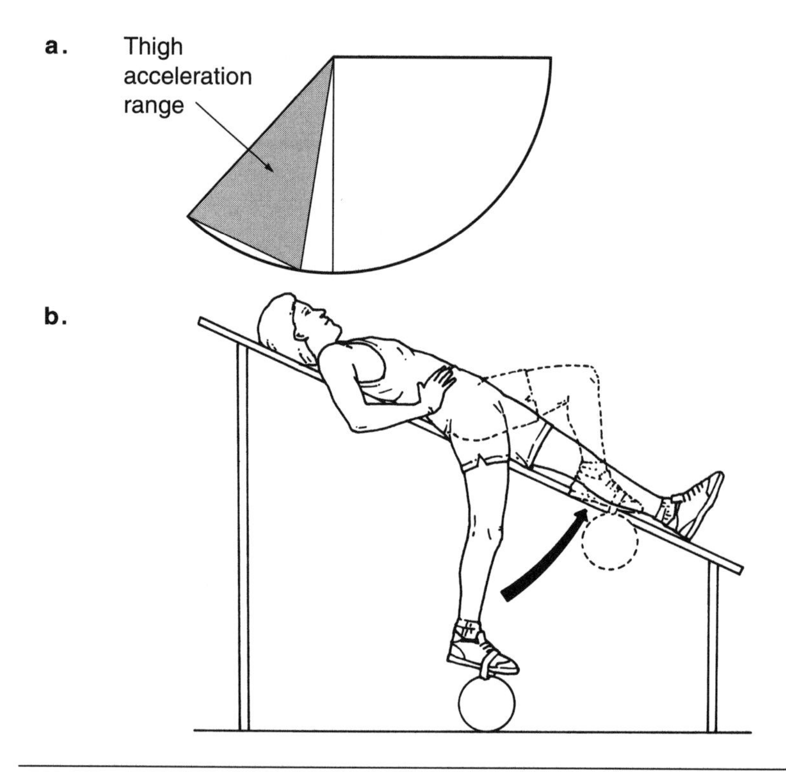

Figure 6.9 (a) "Accentuated" range of motion in swing movement of the leg; (b) an exercise designed to satisfy the requirement for "accentuated" muscular efforts. Note. From *Biomechanics* (p. 100) by D.D. Donskoj and V.M. Zatsiorsky, 1979, Moscow: Fizkultura i Sport.

Accentuation is the most popular exercise strategy among superior athletes in the former Soviet Union and Eastern Europe because this approach best satisfies the requirements for exercise specificity. For instance, there is no reason to develop the strength of hip flexor muscles in their weakest position, as recommended by the peak-contraction principle, because in this range of motion it is the hip extensors, not flexors that are active (recall Figure 6.9). The same is true for exercises with accommodating resistance. There is no need for athletes to train maximal strength over the full range of motion if the maximal force is required in only a small part of the range.

Dynamic exercises that satisfy the requirements for exercise specificity constitute the greater portion of training protocols for qualified and superior athletes. In the strength conditioning of elite soviet athletes in track and field and other dryland summer sports in 1987 and 1988, more than 95% of all sets used free weights or the body weight as resistance. In aquatic sports (swimming, rowing) the proportion of sets with free weights was below 40%.

■ *Squatting or Semisquatting?*

A conditioning coach recommended exercises for the enhancement of leg extensor strength for six groups of athletes—elite, intermediate, and beginning volleyball players and ski jumpers. The elite and intermediate athletes had proper weight training experience, including squatting. The beginning athletes were only slightly accustomed to these exercises. The exercises the coach considered were squatting with a barbell, semisquatting, leg press against a weight, and leg press against isokinetic resistance. She then analyzed the following pros and cons:

- Exercise specificity. Ski jumpers perform takeoffs from a deep squat position; volleyball players almost never jump for height from deep squats.
- Force-posture relationships. Athletes are able to lift a greater load using a semisquat versus a squat technique. For instance, an athlete may be able to lift a 1 body weight (BW) barbell using the squat and to lift a load of 1.2 BW in a semisquat. When full squats are performed, the top effort is required only when knees are deeply bent. However, at the range of joint motion specific to a volleyball takeoff, the leg extension force generated during full squats is far from maximal (the athlete lifts only a 1-BW load but is able to lift a 1.2-BW barbell).

Thus, if the coach favored the peak-contraction principle, she would most probably recommend the deep squats (since the highest requirements are for force production in the deepest knee-bent posture where the potential for force generation is minimal). If she selected the accommodation-resistance approach, the proper exercise would be a leg press against isokinetically adjusted resistance. Finally, if exercise specificity was a matter of primary importance and she favored the accentuation principle, the selected exercises would vary between the two sports. The semisquats would be more specific for volleyball players, while the squats would be specific for ski jumpers.

- Load imposed on lumbar spine and injury risk. These are highest in semisquatting (because of the extremely high load), average in squats, and minimal in leg press.

After consideration, the coach recommended the following (percentage of sets):

Skill level	Volleyball players	Ski jumpers
Elite	60% semisquats 25% squats 15% leg press (against a weight)	20% semisquats 50% squats 30% leg press (against a weight)
Intermediate	30% semisquats 40% squats 30% leg press (against a weight)	10% semisquats 50% squats 40% leg press (against a weight)
Beginning	0% semisquats 25% squats 75% leg press (40% against a weight and 35% against isokinetic resistance	0% semisquats 25% squats 75% leg press (40% against a weight and 35% against isokinetic resistance

In the beginners' group, the weight lifted in deep squats was relatively low (6–10 RM) and primary attention was given to proper lifting technique.

Additional Types of Strength Exercises

Superior athletes mainly use dynamic training exercises of concentric muscle action. Other types of exercises are used in training routines, however, either as supplementary training or for developing specific strength abilities other than F_{mm}.

Isometric Exercises

Isometric training requires no expensive equipment, can be performed anywhere, and, if the number of trained postures is few, takes little time. In spite of these advantages, isometric exercises are used in athletic training mainly as a supplemental training tool, for several reasons. First, they lack the specificity necessary for strength gains (especially for dynamic sport movements). Second, there is little transfer of training effects from the angular position selected for training to other joint angle positions. If a muscle group is overloaded (for instance at 100°), the strength gain will occur at that angle with little improvement at other angles. In addition, these exercises are sometimes painful for superior athletes. The forces isometrically developed by elite athletes are extremely high. In the isometric imitation of lifting a barbell from the floor, for example, the maximal force F_{mm} in the most favorable body position may be well above 8,000 N in elite weight lifters. The mechanical load acting on different body structures, such as the lumbar spine, may exceed safe levels.

A coach who is planning isometric training should keep in mind that accommodation to isometric exercises occurs very quickly. In qualified athletes, strength gains peak out in about 6 to 8 weeks. Thus, the isometric training routine should be planned, maximally, for one to two mesocycles.

The following guidelines govern isometric training protocol:

- *Intensity*—maximal effort
- *Effort duration*—5 to 6 s
- *Rest intervals*—approximately 1 min if only small muscle groups, such as calf muscles, are activated; up to 3 min for large, proximally located muscles
- *Number of repetitions*—usually three to five for each body position
- *Training frequency*—four to six times a week with the objective to increase F_{mm}; two times a week for maintenance of the strength gain
- *Body position*—(a) in the weakest point of the strength curve, or (b) throughout the complete range of motion with intervals of 20° to 30°, or (c) in an accentuated range of angular motion

The second variant on body position is time consuming because many angles within the range must be strengthened. Qualified athletes typically recognize the third variant as the most efficient.

Isometric efforts of large, proximally located muscles may produce a high rise in blood pressure. Individuals at risk of cardiac disease, atherosclerosis, or high blood pressure should avoid these exercises. Athletes should check arterial pressure at least once a week during periods of isometric training.

Because of rapid accommodation, the strength gain from isometric exercises is generally less than from dynamic exercises. This should be taken into account when isometric strength gain is the training objective. A typical example is the "cross," a ring exercise performed in men's gymnastics. As a routine sequence in this case, the gymnast should use dynamic exercises at the beginning (to speed up strength enhancement) and then intermittently add isometrics to improve the specific coordination pattern.

Isometrics are also used in sport to enhance static muscular endurance, for example in long-distance speed skating, where the demand for maintaining a bent trunk posture is extremely high. In 10,000-m skating, the load of the inclined body position has to be sustained for about 15 min.

Isometrics may be used also to improve posture stability, such as that required in shooting a handgun. Holding a 3- to 5-kg weight (instead of a pistol) up to 1 min in the shooting position is a useful training exercise for shooters at the intermediate, not the superior, level. This exercise helps reduce the amplitude of arm microvibration (supposedly by increasing strength in the slow tonic muscle fibers).

Self-Resistance Exercises

Exercises based on self-resistance, not included in the classifications considered earlier, are rarely used in athletic training and are not recommended. In such exercises, the tension of antagonist muscles resists tension of the primary agonistic muscle group. If the muscles are near maximal activation, the training load is extremely high. Healthy people may do these exercises, though cautiously, for general muscle development. Immediately after self-resistance exercising, the muscles become tough and nonelastic (their resistance to palpation or indentation increases) and the circumference of the extremity enlarges. This creates the visual impression of muscular hypertrophy; for this reason, some bodybuilders do self-resistance exercises just before a contest to improve their outward appearance.

The intentional, forced activation of antagonistic muscles, however, harms the proper coordination pattern desired in almost all sport skills. Therefore, self-resistance exercises are not recommended for athletes.

Yielding Exercises

Heavy resistance exercises with eccentric muscle action (*yielding exercises*) are seldom used in strength training. (The term *pliometrics* for these exercises is problematic because of its ambiguity. Strictly speaking, pliometrics refers to movements with eccentric muscle action. However, many authors have used this term for exercises with reversible muscle action, such as depth jumping, in which both eccentric and concentric types of muscle action are involved.)

Eccentric exercises easily provoke *delayed muscle soreness*. All athletes, at one time or another, experience delayed muscle pain, soreness, and a concomitant decrease in strength after exercise sessions. The soreness occurs typically 24 to 48 hr after the workout. Greater soreness is reported with yielding exercises (Figure 6.10).

Several theories of delayed muscle soreness have been suggested. They can be divided into two main groups. The *damage theory* suggests that muscular soreness is induced by damage done to the muscle or connective tissues during exercise. According to the *spasm theory*, on the other hand, a cyclical three-stage process causes delayed muscle soreness. First,

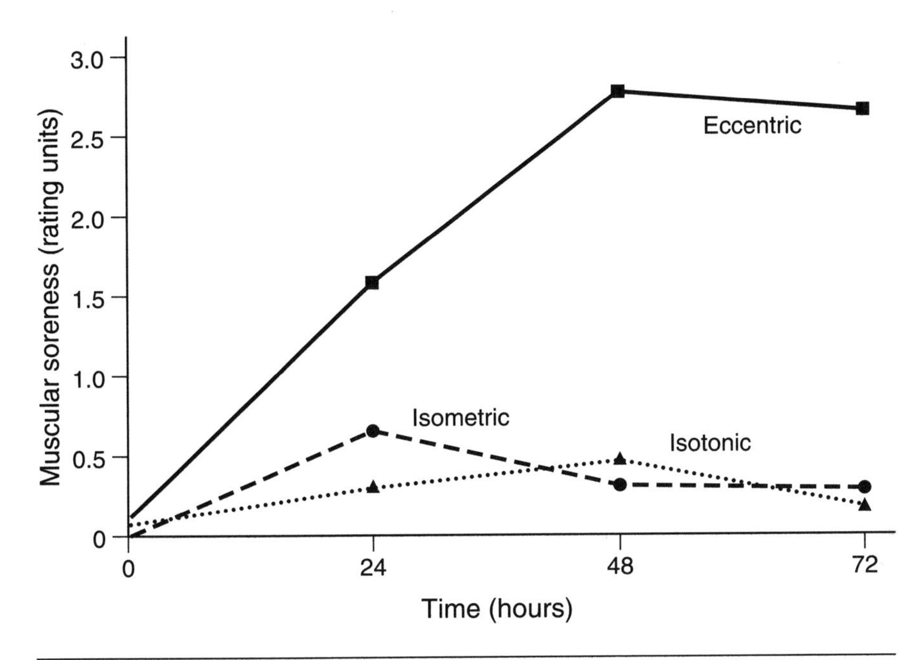

Figure 6.10 Delayed muscle soreness following different training routines. Soreness is most pronounced after yielding exercises. Note. From "Residual Muscular Soreness as Influenced by Concentric, Eccentric, and Static Contractions" by T.S. Talag, 1973, *The Research Quarterly*, **44**(4), pp. 458-469. Copyright 1973 by American Alliance for Health, Physical Education, Recreation, and Dance. Reprinted by permission.

exercise induces ischemia within the muscles. As a consequence of the ischemia, an unknown "pain substance" is accumulated. In turn, the pain elicits a reflectoric muscle spasm. Due to the spasm, ischemia increases, and so forth, and the whole process is repeated in this cyclical manner.

Delayed muscle soreness can be prevented by gradually increasing training intensity and volume. Stretching exercises, especially static ones, are useful for both preventing soreness and reducing its symptoms. Some authors suggest taking vitamin C (100 mg/day, twice the recommended daily dose).

Suggestions about the use of yielding exercises depend on the training objective (i.e., whether the target is concentric, eccentric, or reversible muscle action). When the goal is *concentric* or *isometric* muscle action, exercising with eccentric actions offers no particular advantage. However, an athlete may use these exercises to prepare, for the most part psychologically, for loads above 1 RM. A barbell of very high weight (about 110% of the 1 RM in the relevant movement, e.g., front squat) is actively lowered in these exercises. To prevent accidents, the athlete should be assisted. When these exercises were performed by members of the U.S.S.R. National weight-lifting team in the 1984 to 1988 Olympic cycle, the training volume (the total number of repetitions multiplied by the average weight) did not exceed 1% of the total weight lifted.

Yielding exercises are broadly used in gymnastics for training such stunts as the "cross" on the rings or a horizontal handstand on the parallel bars. Concentric exercises are more efficient for this purpose. However, if used by athletes who are not strong enough to perform these stunts properly, special technical devices or individual assistance is usually required.

Theoretically, eccentric exercises should be used to train the *yielding strength* manifested during landing in parachuting, ski jumping, figure skating, or gymnastics. In these exercises (landing from a large drop distance), however, high impact forces are almost unavoidable, and special precautions must be taken to prevent injury and muscular soreness (exercises should be brought into training gradually; soft surfaces such as gymnastic mats should be used to absorb the impact). It is especially important to perform landings "softly," preventing the heels from hitting the ground. In spite of these precautions, the risk of injury and degenerative changes in articular cartilages and subchondral bones is still too high, so the number of landings should be minimal. Both coaches and athletes need to recognize that overuse as well as inappropriate use of yielding exercises is unsafe.

Yielding exercises should *not* be used for training *reversible (stretch-shortening) muscle action*. The very essence of the stretch-shortening cycle is the *immediate* use of enhanced force production, induced by the prestretch, in the push-off phase. The pause between eccentric and concentric phases of a movement eliminates any advantage that could be gained from the stretch-shortening cycle. This cycle is one uninterrupted movement, not two combined movements. Athletes trained in the landing, rather than the immediate takeoff phase, stop themselves in the lower body position and perform the stretch-shortening action as two sequential movements instead of one continuous movement. Because of the negative transfer of training effect, the use of yielding exercises does not improve performance in the reversible muscle action.

■ "Bouncing" not "Sticking"—Don't Repeat This Mistake!

A coach, trying to improve explosive strength of athletes, advised them to perform, as he said, "pliometrics drills": drop jumps from height in a standing landing posture. The height was from 150 to 250 cm. The landing was performed on gymnastics mats. Although the athletes experienced substantial muscle soreness after the first training session, the coach insisted on continuing the exercise, assuring the athletes that "gain without pain" is not possible. However, in spite of many efforts, the athletic performances in takeoff-related activities did not improve. Moreover, the coordination pattern of the support phase (in jumping or even running) deteriorated. The athletes began to break one uninterrupted eccentric-concentric movement (landing-takeoff, stretch-shortening cycle) into two slightly connected motions: landing and then takeoff.

During natural movements, the primary requirement for a proper motion pattern is not to resist the external force and decrease the body's kinetic energy but to increase the potential for the ensuing takeoff. This goal is realized if both the potential energy of muscle-tendon elastic deformation and the enhanced muscle activation (induced by the interplay of the stretch-reflex and Golgi tendon reflex) are used during the second phase of the support period. If an athlete stops after landing, the potential elastic energy dissipates into heat, and the potentiated muscle activity vanishes. The splitting of one continuous landing-takeoff motion into two motor patterns is a typical "bad habit." This so-called "skill" is very stiff, and much time and effort is needed to correct this mistake. "Bouncing" rather than "sticking" should be accentuated in landing drills.

Exercises with Reversible Muscle Action

In reversible muscle action exercises, a muscle group is stretched immediately before shortening. One example is drop jumping, that is, dropping to the floor from an elevation and then immediately jumping for height. In exercises with reversible muscle action, resistance is determined by the kinetic energy of the falling body rather than its weight (mass) of velocity alone. The kinetic energy (E) is defined by the formula $E = mV^2/2$, where m is mass and V is velocity. In reversible movements (exercises), the same magnitude of kinetic energy can be achieved with different combinations of velocity (dropping distance) and mass. An increase in mass always leads to a decrease in rebound velocity. The moderate increase of velocity at approach initially leads to an increase of rebound velocity, but if the approaching velocity is too high, the rebound velocity decreases (Figure 6.11). The optimal magnitude in approaching velocity (and kinetic energy) depends on the mass of the moving body.

The most popular exercises involving reversible muscle action are single-leg, double-leg, and alternate-step hopping. Among experienced athletes, drop or depth jumping is popular. Many coaches assume that drop jumps are directed toward improving the storage and reuse of elastic

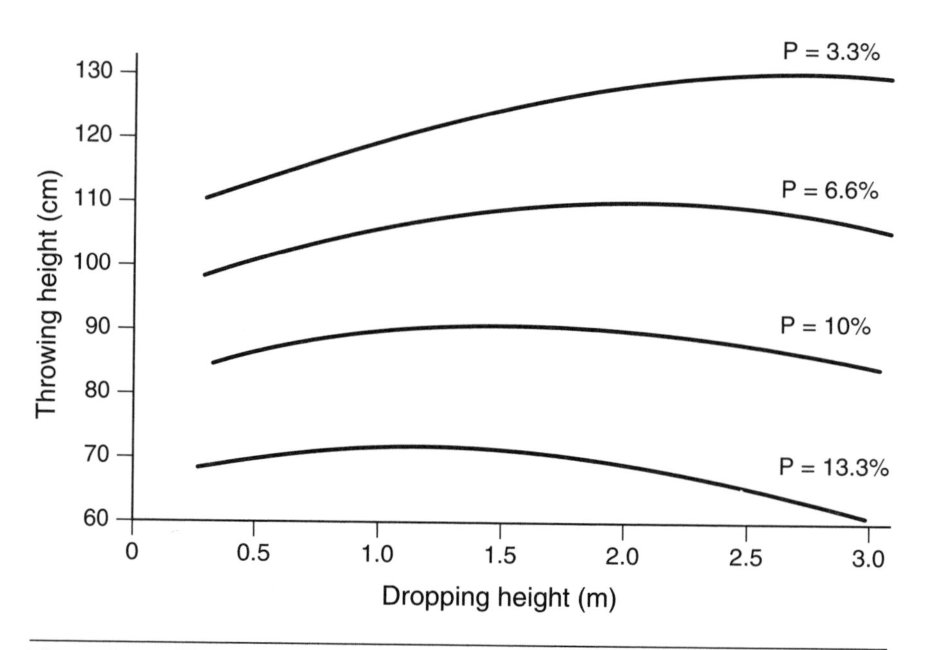

Figure 6.11 Changes in the height of an implement thrown as a function of its weight and dropping height. With a special installation in laboratory conditions, shots of different weights (3.3; 6.6; 10; 13.3% of F_{mm}) were dropped from various heights (from 0.5 to 3.0 m). An elite shot-putter put shots vertically (in supine position). The height of each throw was measured. According to basic mechanics, dropping height is proportional to the approaching velocity squared and the throwing height is proportional to the squared rebound velocity. Note. From *Special Strength Training in Sport* (p. 145) by Yu.V. Verchoshansky, 1977, Moscow: Fizkultura i Sport. Reprinted by permission from the author.

energy during takeoff. However, more energy is stored and reutilized only if the muscle tension is greater (see chapter 2). So the actual source of enhanced motor output is the increased muscle force production during this type of activity. The enhanced force is a result of

- inhibition of the reflex from Golgi organs (because this reflex is inhibitory, an "inhibition of the inhibition" takes place),
- potentiation of the stretch reflex, and
- proper timing.

There are two main variations on the drop jump. It is done with small amplitude of leg flexion during landing and takeoff ("bouncing") or large-amplitude flexion ("squatting" or "countermovement jump").

Bouncing should be performed with minimal contact time. Athletes are advised to do the takeoff as though the surface is hot like a frying pan.

The dropping distance should be adjusted to keep the athlete's heels from hitting the ground. The horizontal velocity at landing should be high enough to avoid plantar hyperflexion. Squatting is recommended to improve jumping ability in vertical jumps (e.g., for basketball, volleyball); starting velocity in football, ice hockey, and sprinting; and the explosive strength of football linemen, throwers, and weight lifters. Squats, though, should not be too deep. The range of knee flexion should be only slightly greater than in the primary sport movement.

Typically, a jumper makes initial ground contact with extended legs. However, if the aim is to improve the rate of force development, especially in the knee extensors, exercises requiring landing on a bent leg may be used. This is the case also when the athlete wants to improve landing on a flexed leg (for instance, figure skaters, while performing jumps with several twists, land on a flexed support leg).

Practical experience shows that dropping jumps are a very effective drill. However, the injury risk is high and accommodation to these exercises occurs very quickly. Therefore, these guidelines are recommended:

1. Follow the prescribed sequence of exercises during multiyear training—regular jumping exercises, weight training exercises, and then drop jumping. Drop jumps should not be used by young athletes with training experience of less than 3 to 4 years.
2. Do not use drop jumps continuously for more than one or two mesocycles. Vary exercises by performing with and without additional weight vests (belts). After initial adaptation (usually two to three training workouts), use weight vests for 2 to 3 weeks; then exercise without weights and increase the dropping distance gradually.
3. Maintain the proper level of explosive strength during the competition period by doing drop jumps once every 7 to 10 days. Exclude these jumps from the training program at least 10 days before an important competition.
4. Determine the exercise intensity (kinetic energy, weight, dropping distance) on an individual basis. The main requirement is proper technique (i.e., smooth transition from the yielding phase to the push off, heels not hitting the ground).

Drills for training the stretch-shortening cycle should not be limited to drop jumps, though often they are. The possibility of increasing the mass of the falling body is rather limited in drop jumps—people wear weight vests or belts, but these cannot be as heavy, for example, as 100 kg. In view of the complex relationship between kinetic energy, velocity, and body mass, on the one hand, and the motor output of reversible muscle action, on the other, training with stretch-shortening cycle devices, where both the mass and velocity may be changed, is recommended. An example of such a machine is shown in Figure 6.12.

Figure 6.12 A "swing" exercise machine for training reversible muscle action in landings and takeoffs. Both the range of motion and mass of the system are varied in training. The mass of the swinging assembly may be increased up to 200 to 300 kg and even more; this is important when training qualified athletes.

Sport Exercises with Added Resistance

You can best meet requirements for exercise specificity when you use the main sport movement, with increased resistance, for training. Examples are uphill cycling and cycling with a changed gear ratio.

Each sport event is performed against a given resistance and a given velocity. The resistance is predetermined by the mass of the implement or the athlete's body (inertia forces) and by body dimensions (aerodynamic or hydrodynamic forces). If an athlete performs the movement as fast as possible, movement velocity is a function of the resistance (an additional example of a parametric relationship; see chapter 2). If the resistance increases, the velocity decreases.

Resistance in *terrestrial* athletic events can be increased by adding weight, by uphill movement, by slowing progression, and by increasing aerodynamic resistance with parachutes.

Implements of heavier weight such as weight vests, belts, wrist cuffs or ankle cuffs may be worn. Although adding this weight is simple, note that it is principally the demand for vertical force (acting against gravity) that is increased with supplementary weight. The typical requirement in athletics is to increase the horizontal component of the exerted force. Exercising with additional weight requires that force be exerted in an inappropriate (vertical) direction. In running, for instance, this leads to excessive body lift in the flight phase. Furthermore, locomotion using additional weights, especially ankle weights, increases impact stresses on lower extremities.

Including some form of *uphill ambulation*, such as running, walking, or skiing, is limited by the possible changes in sport technique. Some coaches have tried *retarding the athlete's progression*. For example, athletes run with a harness, tow a sled, or use a pulley machine with a weight stack. These methods are cumbersome in that the equipment is bulky and heavy. Typically they are used only in short movement ranges (e.g., for the sprint start, but not for sprint running).

Increased aerodynamic resistance, on the other hand, is a popular method among elite athletes in sports such as speed skating and running. Small parachutes are used for this purpose (Figure 6.13). When the athlete runs, the parachute inflates, creating a drag force. The higher the running velocity, the greater the resistance force. Parachutes of several different sizes are used in training. The impeding drag force, depending on parachute size, may vary from about 5 to 200 N (within the speed range 6 to 10 m/s).

Parachutes offer several advantages over other methods of resistance training:

- The resistance (drag) force acts strictly in the direction of the athlete's movement;
- Sport technique is not negatively altered;
- Parachutes are not limited to use in straight ambulation, but can also be used when the athlete is running curves, running over hurdles, or changing direction (e.g., football, soccer);
- Parachutes weigh only a few ounces; and
- A parachute can be released while the person is running; this provides an impetus to increase movement velocity (this is called an assisted drill).

The only drawback of parachutes is that they offer the same amount of resistance in both the support and the nonsupport phases of running. Thus they hamper movement speed during flight while slightly changing the position of body joints during foot landing, as in hurdle running.

For maximum effect, one should vary the parachute size in micro- and mesocycles as well as in workouts. Resistant and customary training are executed during preparatory microcycles, while the assisted drills are mainly utilized near the competition season. In a workout as well as in a sequence of training blocks, the resistance, determined by the parachute

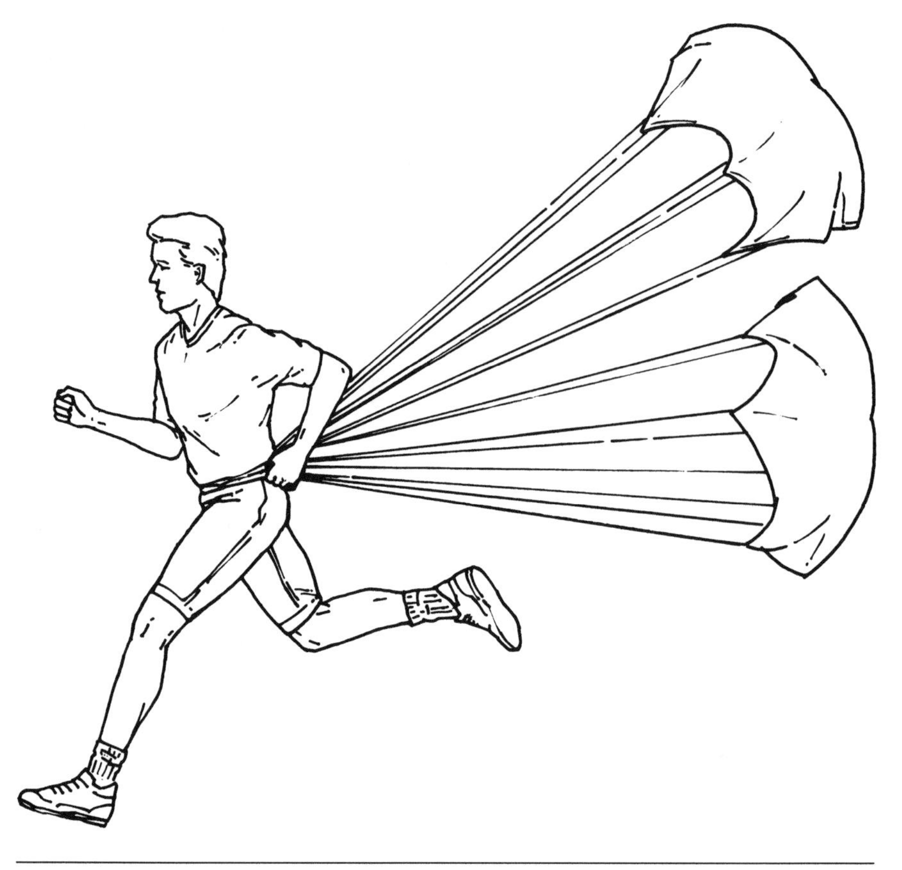

Figure 6.13 Use of a parachute in running drills.

size, is decreased by degrees. During a training workout, first drills (after warm-up, naturally) are performed under the heaviest resistance called for during that training session, and the final attempts are executed under the lightest resistance. Before and immediately after parachute drills, the same drills are performed under normal conditions. Parachutes are typically used two to three times a week. Sessions with parachutes are interspersed with the usual workouts. During a competition period, parachutes are used to induce a feeling of enhanced speed and explosiveness. For contrast, they are used three to five times within sport-specific drills at the beginning of a session, followed by the usual drills without a parachute.

In *aquatic* sports such as swimming or rowing, hydrodynamic resistance can be increased. With this objective, the streamlining of the body or its frontal area is altered. This can be accomplished by increasing the resistance offered by the boat or the swimmer's body or by expanding the hydrodynamic resistance of the propeller (the blade of the oar in rowing, the paddle in kayaking and canoeing, the swimmer's arm), for instance, using hand paddles is common in swimming.

In both cases, the force exerted by the athlete increases. However, the mechanisms of the force output augmentation are biomechanically different, so the training effects are also dissimilar. In aquatic locomotion, the external force developed by an athlete is determined by both athlete strength, in particular the individual's force-velocity curve (parametric F_m-V_m relationship), and the water resistance offered (Figure 6.14). As in all parametric relationships, force decreases as movement velocity increases. An athlete cannot develop high force at a high velocity of muscle shortening. Conversely, water resistance increases with a gain in velocity. Note that, in the first case, velocity is relative to the athlete's body (in essence, it is the velocity of muscle shortening); in the second case, the velocity of the propeller, relative to the water, is the point of interest.

The exerted force is indicated in the third panel of the figure by the bold arrow. To the left of this point, where velocity is small, the athlete's strength is higher than the hydrodynamic resistance. Picture an athlete slowly moving his or her arm or paddle in the water. No matter how strong the person is, the exerted force is limited by water resistance, which is low in this case due to low velocity. However, if the movement

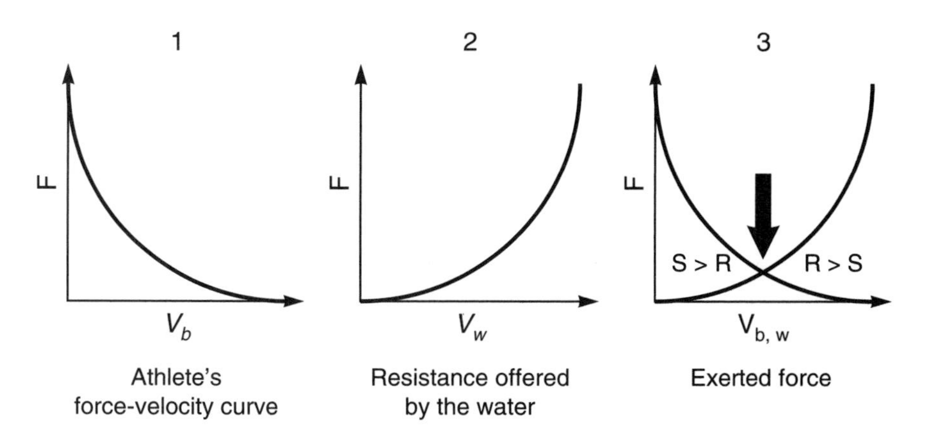

Figure 6.14 The force exerted by an athlete is determined by the interplay of (1) the force-velocity curve (the maximal force developed by an athlete at a given velocity when high resistance is met) and (2) water resistance. The interception of these curves corresponds to the force exerted by the athlete against water resistance (3). To the left of this point the strength ("potential") of the athlete exceeds the amount of resistance (S > R); to the right, R > S (S, strength; R, resistance). V_b is relative velocity of body segments; V_w is velocity of the body relative to the water.

velocity and the corresponding water resistance are high enough, the demand for a large force can exceed the athlete's capacity. In this case, the athlete's ability to produce sufficient force is the limiting factor.

It is known that, biomechanically, a propeller's velocity relative to the water in the direction of the boat's (body's) motion (V^p_w) equals the difference between the velocity of the propeller with respect to the boat (V^p_b) and the amount of boat (body) velocity (V^b_w):

$$V^p_w = V^p_b - V^b_w$$

When hydrodynamic resistance of the boat or the swimmer's body increases, the boat's (body's) velocity relative to the water (V^b_w) decreases. Furthermore, if propeller velocity relative to the boat (body) is kept the same, then its velocity relative to the water (V^p_w) increases. So, when stroking at the same velocity with respect to the boat or body (V^p_b = Constant), the athlete meets greater water resistance by virtue of the *increased* propeller velocity relative to the water (V^p_w).

When the hydrodynamic resistance of the propeller is increased (e.g., with hand paddles), the same stroking speed (V^p_b) produces greater body (boat) velocity (V^b_w). The propeller's velocity relative to the water (V^p_w) then decreases instead of increasing as in the previous case. The exerted force, however, increases as a result of poor streamlining of the propeller (Table 6.1).

It is recommended that these additional resistances be raised alternately. Note also that the amount of added resistance (exercise intensity) is limited by change in sport technique. If the technique is altered substantially, the intensity (i.e., aero- or hydrodynamic resistance) must be decreased.

Table 6.1 Boat (Body) Versus Propeller (Paddle, Hand) Resistance Changes

Resistance increased	Velocity of the propeller relative to the body, V^p_b	Velocity of the body relative to the water, V^b_w	Velocity of the propeller relative to the water, V^p_w	Exerted force is greater due to the greater...
Boat (body)	=	<	>	velocity, V^p_w
Propeller	=	>	<	resistance offered

Electrostimulation

During the last two decades, many attempts have been made to use transcutaneous muscle electrostimulation (EMS) as a training method for athletes. The method was originally developed in the former Soviet Union in the late 1960s.

In theory, one advantage of EMS is the activation of predominantly fast motor fibers that are difficult to recruit voluntarily. During EMS, the size principle of motor unit recruitment is no longer valid; fast-twitch motor fibers are activated first in this case. These have a lower threshold to externally applied electric current and, in addition, many are located superficially, close to the external edge of muscles.

EMS is a useful supplement to conventional strength training methods (Figure 6.15). It can enhance not only maximal stimulated force but also voluntary force, speed of motion, and muscular endurance. The time to accommodation is usually about 20 to 25 training days in conditioning for maximal strength and 10 to 12 days for maximal velocity. During EMS training for muscular endurance, the leveling off is not attained even after 35 sessions. Positive results, including improvement in sport performance, have been demonstrated in weight lifting, gymnastics, and track and field events, as well as in the jumping ability of volleyball and basketball players.

However, contrary to popular opinion, soviet and Russian athletes have not regularly used EMS as a substitute for traditional strength training. Athletes' attitudes toward this method vary substantially. Many elite athletes have been very positive regarding EMS use. For instance, some Olympic champions in kayaking and canoeing have sought to stimulate several muscles, including the biceps brachii and deltoid, over a 1-month period before important competitions, including the Olympic games.

At the same time, in spite of evidence that maximal strength may be enhanced as a result of EMS, this method has not been accepted by numerous qualified athletes. In addition to a customary conservatism, there are two main reasons. First, athletes cannot use enhanced isometric (specially stimulated) values in real sport events. The time and effort needed to transmute acquired changes into force output of the real movement is too great. Second, some athletes using EMS have an unpleasant feeling of lack of muscular control and a loss of coordination and simply refuse to continue. These findings confirm the idea that, loosely expressed, only muscles (not neural factors) are trained with EMS. The ability to activate trained skeletal muscles does not seem to be augmented as a result of this kind of protocol.

There may be several reasons for such different attitudes on the part of athletes toward EMS. First, EMS can be used in improper proportion to conventional strength training. If the proportion of EMS training is too great (for a given athlete), the transmutation may become difficult. And second, inap-

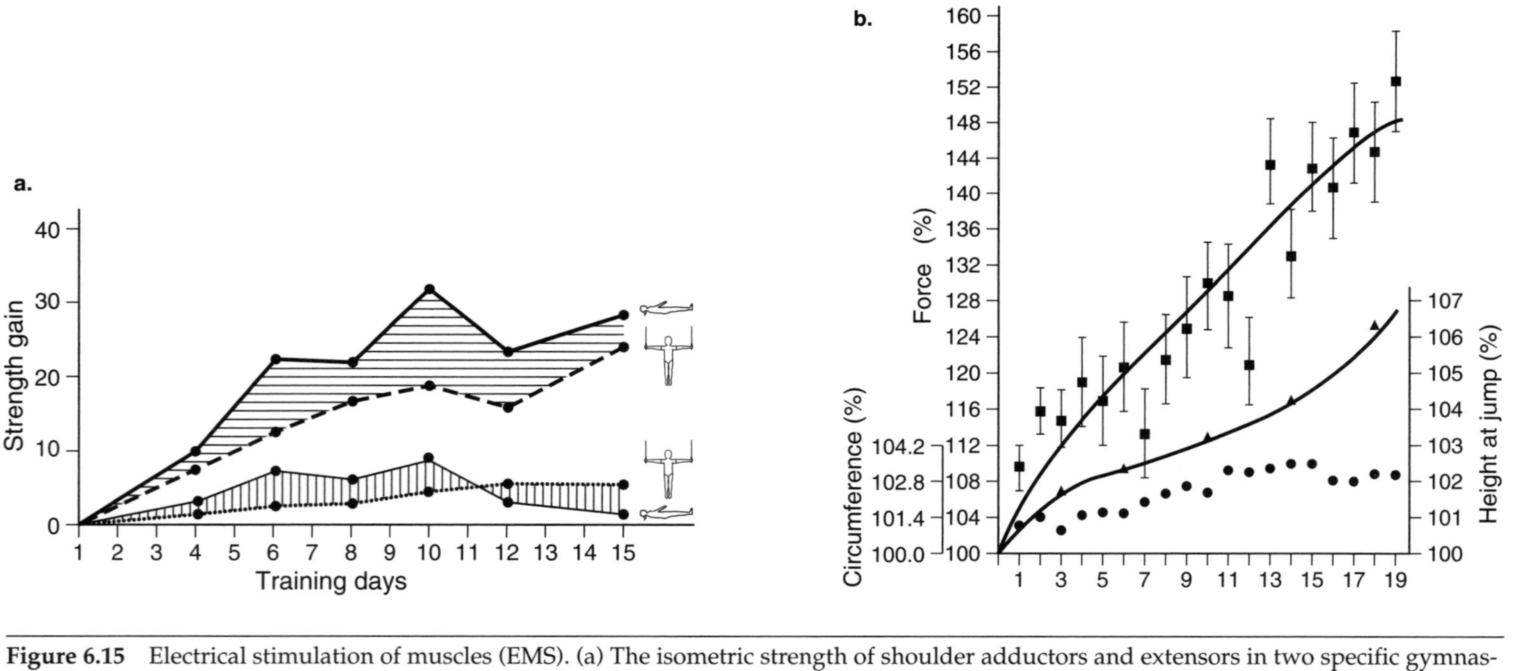

Figure 6.15 Electrical stimulation of muscles (EMS). (a) The isometric strength of shoulder adductors and extensors in two specific gymnastic postures. National soviet Olympic men's gymnastics team (average data). Abscissa, training days; ordinate, strength gain, % of F_{mm} in corresponding posture. Two upper curves, an experimental (EMS) group; two lower curves, the control group. Note. Adapted from *Muscle Electrostimulation in Training of National Men's Gymnastics Team* (p. 36) by Y.M. Kots and E.S. Belov, 1973, Moscow: Central Institute of Physical Culture. Reprinted by permission from the authors. (b) EMS of calf muscles (m. triceps surae). Relative gains (% of initial values) are shown: (a) maximal strength of plantar flexion (black rectangles), (b) shank circumference (circles), and (c) standing jump height (triangles). Eight subjects, ages 16 to 17 years, bilateral EMS training. Note. Adapted from "Muscle Strengthening via Electrical Stimulation" by Y.M. Kots and V.A. Chvilon, 1971, *Theory and Practice of Physical Culture* (4), pp. 66-72. Reprinted by permission from the authors.

propriate muscle groups can be selected for the EMS training (again for a certain athlete). If the strongest muscle from the muscle group is stimulated and the weakest one is not, there is no performance improvement.

To date, EMS has been routinely used by qualified athletes in only isolated cases. One example is correction of the "junctional flatfoot," an acute arch flatness of the foot occurring as a result of high training loads in runners and jumpers. Regular EMS (twice a day) of the small arch muscles helps prevent and treat this malady. Another example is stimulation of the spine erectors in athletes, in particular rowers and kayakers, who are susceptible to low-back pain. EMS is also used in training of the shoulder adductor muscles used to perform the cross in men's gymnastics.

The following EMS routine, known as the Russian protocol, is typically used:

- Carrier signal—sinusoidal or triangle
- Frequency—above 2,500 Hz
- Modulation—50 Hz
- Duty cycle—50% (the signal is applied for 10 ms with a 10-ms interval between trials)
- Stimulus amplitude (SA)—adjusted individually to induce a force above 100% of maximal voluntary isometric force F_{mm} or to the limit of subject tolerance; SA depends on the output impedance of the stimulator and typically exceeds 90 volts
- Contraction time—10 s
- Rest between contractions—50 s
- Number of contractions—10 per day
- Number of training days—5 per week

The most important feature of the described stimulation protocol is the frequency of the carrier signal, which should be located in the sound frequency band, i.e., above 1,500 Hz. EMS, when performed properly, is painless. The electrode surface must be wetted with a special paste to achieve homogeneous electrical resistance at the skin-electrode interface.

Present knowledge about EMS is not satisfactory as a basis for a final recommendation. The prospects for using EMS in athletic training should be further investigated.

Breathing During Strength Exercises

If maximal force is exerted while inhaling, exhaling, or making an expiratory effort with the glottis (the opening between the vocal cords) closed (the last is called the *Valsalva maneuver*), the amount of force developed increases from inspiration to expiration to Valsalva maneuver. The underlying mechanism for this phenomenon is a pneumomuscular reflex, in which increased intralung pressure serves as a stimulus for the potentiation of muscle excitability. The true mechanisms of enhanced muscle excitability have yet to be studied.

Thus, although the Valsalva maneuver might be considered a useful breathing technique for ultimate force production, it also provokes a cardiovascular response that many physicians consider harmful, particularly in individuals with heart problems. Because air cannot escape, the intrathoracic pressure sharply increases (up to 40–100 mmHg and even higher, whereas normally it is 2–15 mmHg lower than atmospheric pressure). Because of the high intrathoracic pressure and associated compression of the venae cavae, which return blood to the heart, venous return to the heart decreases. In turn, both stroke volume and cardiac output decrease. As a result of the small venous return and high intrathoracic pressure, the heart dimensions, particularly the chamber dimensions, are lessened (this is called the Valsalva effect). The decreased stroke volume is compensated for by increased heart rate, sometimes above 170 beats per minute. In addition, blood pressure increases substantially. (Values up to 320/250 mmHg have been measured during barbell squats.) The elevation is explained mainly by the high intramuscular pressure, which results in increased total peripheral resistance and increased blood pressure.

The decreased cardiac output may further result in brain anemia and a loss of consciousness. (This has happened many times during weight-lifting competitions involving the military press; this lift has been excluded from the Olympic weight-lifting program since 1972.) Immediately after the lift, intrathoracic pressure abruptly falls and a large amount of blood overfills the heart. Then, both stroke volume and cardiac output rise, blood pressure decreases, and after some time, all values return to normal.

However, athletes adapt to such changes, and properly planned and executed strength training does not cause hypertension. Contrary to common misconceptions, heavy resistance training (again, if properly planned and executed) results in positive adaptations of the cardiovascular systems. At the same time, athletes and coaches should exercise these cautions during physically strenuous activities:

1. Permit the Valsalva maneuver, or expiration efforts with a closed glottis, only during short-time ultimate efforts. Beginners often stop breathing during repetitive lifts of low intensity. A coach should discourage this practice. In principle, high intrathoracic pressure is undesirable; on the other hand, high intraabdominal pressure is considered useful. The torque generated by intraabdominal pressure reduces the compressive force acting on the intervertebral disks and may lessen the probability of spinal disk injury and, ultimately, increase lifting ability (see chapter 7).
2. Beginners should not be given many exercises with ultimate and near-ultimate efforts.

3. An athlete should not inspire maximally before a lift. The maximal inhalation unnecessarily increases intrathoracic pressure.
4. Forced expirations, rather than the Valsalva maneuver, should be used whenever possible.
5. Beginners should inhale and exhale during performances, especially when the weight is held on the chest.
6. Finally, there are two ways to match breathing phases (the inspiration and expiration) with the performed movement—the anatomical and the biomechanical match.

This last point requires some elaboration. In movements with small efforts (similar to those in calisthenic exercises such as a trunk inclination) the inhalation should coincide with the trunk extension and the exhalation with the trunk bending. This is called an *anatomical match* (of breathing phases and movement). In contrast, when high forces are generated the expiration should match the forced phase of movement regardless of its direction or anatomical position. For instance, rowers exhale or use the Valsalva maneuver during the stroke phase when the greatest forces are developed; nevertheless, the legs and trunk are extended at this time rather than flexed (as compared with trunk and leg flexion without an additional external load in calisthenic exercises). This breathing is termed a *biomechanical match*. During strength exercises, the breathing phases and movement should be matched biomechanically rather than anatomically.

Summary

Strength exercises are classified in various ways. For example, they may be static (isometric) or dynamic (concentric, eccentric, reversible, isokinetic). They may concentrate on particular muscle groups, whose comparative strengths are called strength topography. Or they may be classified according to how specific they are to the sport task.

With beginners, especially youngsters, strength topography is the main concern in selecting strength training exercises. For example, you should choose the most important muscle groups, strengthening muscle groups that might be at risk for injury if they were weak, training proximally located muscles, and strengthening muscles that are needed to perform sport movements. For more advanced or mature athletes, however, the goal is to select strength training exercises that are specific and mimic the movement pattern used in the actual sport skill. This is a complex demand, however, and requires careful analysis of movement including, resistance, the timing and rate of force development, movement direction, and variations of muscle strength over the range of joint motion.

Muscular strength varies over the full range of joint motion, depending on changes in both muscle lever arms and muscle force. To manage this force-posture relationship, trainees may use the peak-contraction principle, which focuses on increasing muscle strength primarily at the weakest points of the human strength curve through the selection of proper body positioning, special training devices, and a slow starting motion. Or they may develop maximal tension throughout the complete range of motion (accommodating resistance, used with some physical therapy isokinetic equipment and in some training machines). A third method is training in the range of the main sport movement where the demand for high force production is maximal (this accentuation method has been popular with Russian and eastern European athletes).

Isometric exercises are seldom used, and self-resistance and yielding exercises carry risks, so they are not recommended. Exercises with reversible muscle action are effective but they also carry high risks of injury and accommodation occurs quickly with these exercises. Sometimes additional resistance is added to main sport exercises that best meet the requirements for sport specificity.

During the last 25 years, electrostimulation of muscles has become a popular method to enhance muscular strength. Although electrostimulation shows promise for strength training, this method needs further investigation.

Breathing patterns also affect force production, and many people consider certain patterns, such as the Valsalva maneuver, potentially harmful because of the cardiovascular response they can provoke. With small efforts, the inhalation should coincide with the trunk's extension and the inhalation with the trunk's bending (an anatomical match of breathing phases and movement). With high force, however, expiration must match with the forced phase of movement, regardless of the direction or anatomical position (a biomechanical match). During strength exercises the breathing phases and movement should match biomechanically (rather than anatomically).

CHAPTER 7

Injury Prevention

Heavy resistance training is a relatively safe activity, as the incidence of injuries is low. The risk of injury for a well-coached strength training program has been estimated to be about one per 10,000 athlete-exposures. (An athlete-exposure is one athlete taking part in one training workout or competition). Compared to tackle football, alpine skiing, baseball pitching, and even sprint running, strength training is almost free of risk. However, athletes exercising with heavy weights who neglect certain training rules are susceptible to trauma.

Training Rules to Avoid Injury

Common sense and professional knowledge dictate how to avoid injury. The rules are very simple:

- Maintain the weight-lifting room and exercise equipment in proper order.

- Make sure athletes warm up.
- Do not overdose. Do not recommend the maximal effort method for beginning athletes.
- Be cautious with the use of free weights.
- Provide assistance when a barbell weight exceeds maximum weight and yielding exercises are being performed.
- Emphasize harmonic strength topography; avoid imbalance in muscle development (see chapter 8).

In addition, there is one issue in the strength training paradigm that warrants special attention—the lumbar spine region. In the discussion that follows we consider this concern in detail.

According to epidemiological data, up to 80% of the adult population suffer temporary or chronic pain in the low-back region (the so-called low-back pain syndrome; LBPS). LBPS as a cause of inability to work is either first or second among all illnesses, yielding only to flu and catarrhal diseases. In athletes doing strength training, lower-back damage comprises 44% to 50% of all the injuries sustained.

In addition to such factors as metabolic abnormalities, infections, and genetic predisposition, biomechanical factors (especially spine overloading) are regarded as the primary causes of LBPS. However, in spite of the great mechanical load imposed on the lumbar region in sports like weight lifting and rowing, many elite athletes in these sports have no spinal problems during their lives. Proper sport techniques and fundamentally sound training patterns provide reliable protection against LBPS.

Although we do not know the precise cause of LBPS, volumes of data have indirectly shown that changes in intervertebral disks are usually the initial cause of pain.

Biomechanical Properties of Intervertebral Disks

Intervertebral disks consist of a fibrous ring, the annulus fibrosus, and a jellylike nucleus, the nucleus pulposus. In young persons the jellylike nucleus contains up to 85% water, and the laws of hydrostatic pressure apply—namely, Pascal's law stating that pressure is distributed equally on all sides. Intradisk pressure can be determined by inserting a needle with a pressure gauge into the jellylike disk nucleus. With age the water content of intervertebral disks is gradually reduced, and the laws of hydrostatic pressure cease to manifest themselves in the nucleus pulposus.

When disks are loaded in different directions, their mechanical properties are different. When two vertebrae are compressed with the disk connecting them along the axis of the spine (the Y-axis), the hydrostatic pressure in the nucleus is approximately 1.5 times greater than the average pressure acting on the disk surface (we'll designate this F). Here the vertical pressure on the fibrous ring amounts to just 0.5 F. On the other hand, when horizontal pressure occurs, the disk stretches from within and the force reaches 4 to 5 F on the surface of the fibrous ring (Figure 7.1).

The fibrous ring consists of several cylindrical layers, each of which has fibers proceeding at an angle of approximately 30° to the horizontal; but the directions of the fibers' paths change in adjoining layers. In the disks of young and elderly persons, with an identical external mechanical load, both the amount of pressure acting on particular layers of the fibrous ring and its direction are different (Figure 7.2).

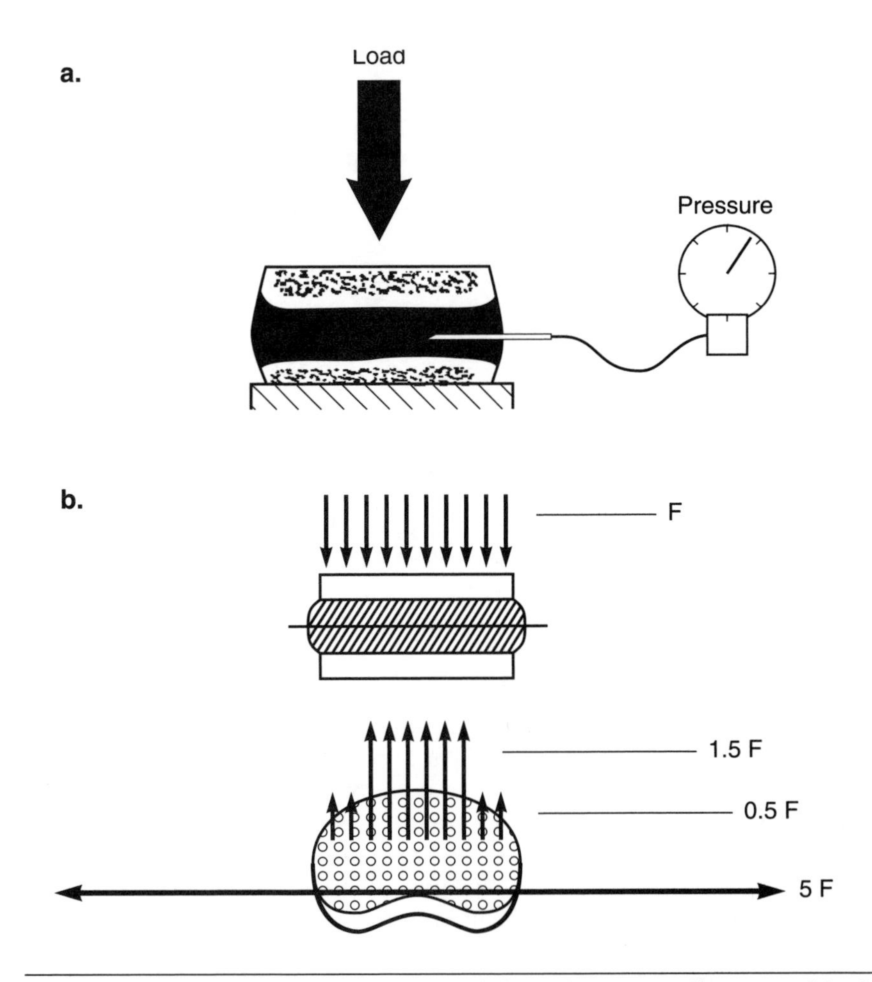

Figure 7.1 Pressure in the intervertebral disks under a vertically imposed load. (a) A scheme of measurement. (b) Pressure distribution. The compressive stress in the nucleus pulposus is 1.5 times higher than the externally applied load (F) per unit area. Note. Adapted from "Towards a Better Understanding of Back Pain: A Review of the Mechanics of the Lumbar Disc" by A. Nachemson, 1975, *Rheumatology and Rehabilitation*, **14**, pp. 129-143.

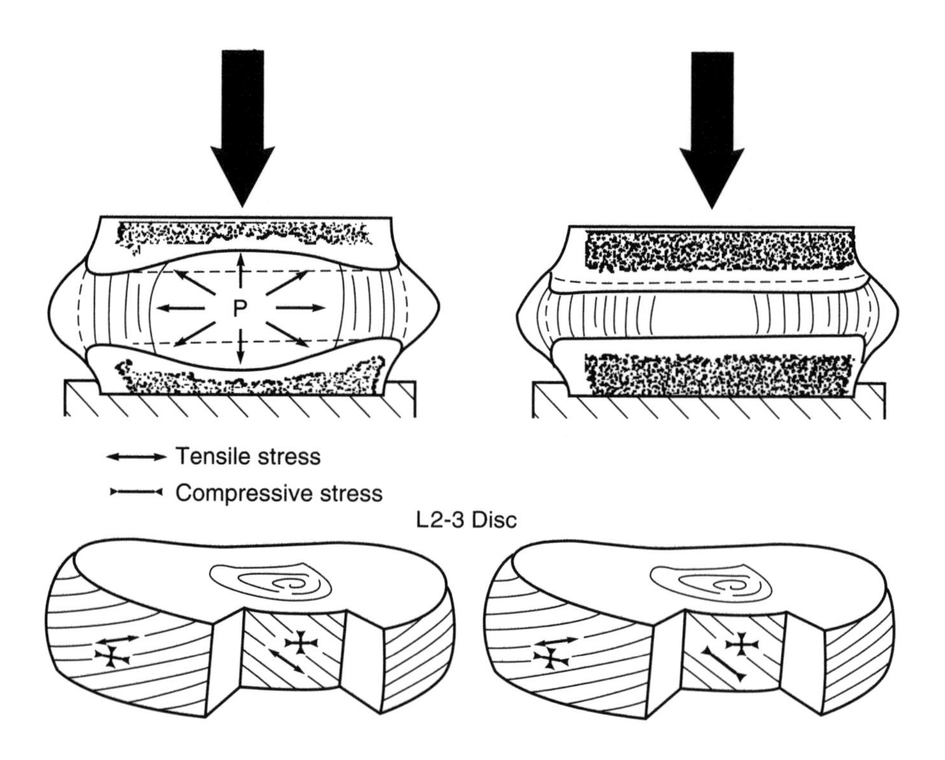

Figure 7.2 Pressure affecting individual layers of the fibrous ring in normal disks (for young persons, left) and degenerated disks (for elderly persons, right). Notice the change in the amount and direction of the pressure. Note. Adapted from "Stress Analysis of the Lumbar Disc-Body Unit in Compression: A Three-Dimensional Nonlinear Finite Element Study" by S.A. Shirazi-Adl, S.C. Shrivastava, and A.M. Ahmed, 1984, *Spine*, **9**, pp. 120-127.

The mechanical strength of disks during a vertical load is adequate; it is not inferior to the strength of adjoining vertebrae. However, a strictly vertical load on the spinal column is not typical for actual everyday situations. Even during regular standing posture, the load does not operate precisely along the axis of the vertebrae (the Y-axis) because of the curvature of the spinal column. It follows from biomechanical analysis that people are the most susceptible to trauma in situations in which a considerable mechanical load affects the intervertebral disks during trunk bending or rotation.

During a lean of the spinal column, the nucleus pulposus is shifted to the side opposite the lean and the fibrous ring is somewhat protruded (Figure 7.3). This may induce compression of the spinal cord rootlets and cause a painful sensation.

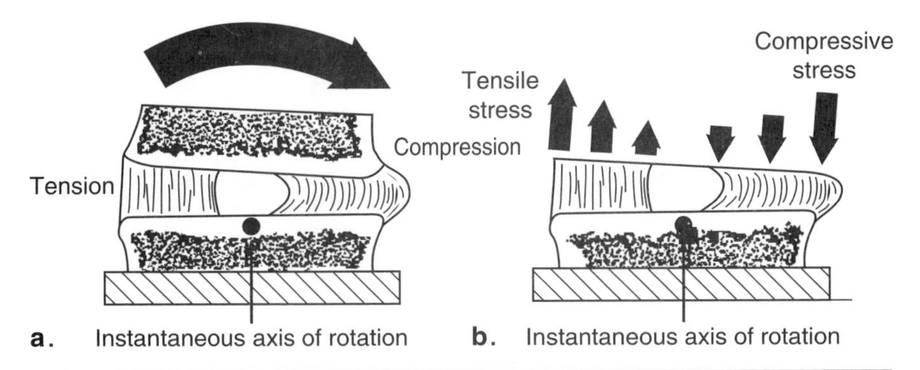

Figure 7.3 (a) Disk deformation; (b) mechanical stresses. Note. From *Clinical Biomechanics of the Spine* (p. 15) by A. White and M.M. Panjabi, 1990 (2nd ed.), Philadelphia: J.B. Lippincott. Copyright 1990 by Augustus A. White & Manohar M. Panjabi. Reprinted by permission.

Mechanical Load Affecting the Intervertebral Disks

Intervertebral disks are affected by impact and by static loads. The latter include loads encountered not only during the maintenance of a given posture but also during the execution of relatively slow movements, when it is possible to ignore waves of impact deformation.

Impact Loads

Landings from gymnastic dismounts, jumping, and running cause the body to undergo an impact load, spreading shock waves to the spine. We can estimate the impact load by the magnitude of acceleration registered on different parts of the body.

In ordinary walking the difference between accelerations of the pelvic region and those of the head amounts to 0.5 to 1.0 g (g is the acceleration due to gravity, g = 9.81 m/s^2). The spine must absorb a shock of similar magnitude with each step. Research on 50-m ski jumping has shown that accelerations of the pelvic region at the moment of landing exceed 10 g; at the same time, the intraabdominal pressure (to be discussed later in this chapter) reaches 90 mmHg. Loads on the spine were reduced when jumpers performed deep (about 40 cm) squats and were increased when they landed with straighter legs. Loads increased in proportion to the sine of the angle between the direction of the speed vector and the slope of the mountain. These examples evidence the exceptionally large loads that the spinal column is subjected to during landings in different sport exercises.

The softening (shock absorption) of an impact load during landing is provided by the combined influence of

- the properties of the supporting surface,
- footwear quality,
- the dampening properties of the motor system, primarily the foot and the knee joints (in persons suffering from LBPS these properties are often reduced), and
- landing techniques.

With "soft" landing techniques, in which ankle plantar flexion and knee flexion are coordinated, the magnitude of impact forces is sharply reduced. During soft landing by experienced athletes, only 0.5% of the body's kinetic energy is spent to deform body tissues (bone, cartilage, spine). During a stiff landing, the deformation energy amounts to 75% of the body's mechanical energy. The difference is 150-fold (75/0.5 = 150).

■ *Land Properly*

In order to prevent spinal injuries during landings, use mats and shoes with good shock-absorbing capacities and employ proper landing motor patterns. Touch the ground with legs extended and feet plantar-flexed, and, immediately after ground contact, avoid a stiff landing by flexing the knees. Practice soft landings, without impact. Good ballet dancers land in such a way that virtually no sound is made! Try to follow this pattern.

Static Load Acting on Intervertebral Disks

Forces that affect intervertebral discs can significantly exceed the body's weight and the weight being lifted. They are produced chiefly by muscle tension. Let's look at the mechanism that causes these loads by examining an example of upright standing posture (Figure 7.4).

Mechanism of Origin

In this case, the weight of the upper body acts on L4 (the fourth lumbar vertebra). The center of gravity of the upper body is not situated directly over the intervertebral disk, but somewhat in front of it. Therefore, a rotational moment of the force of gravity, causing the upper half of the body to lean forward, must be opposed by a counterbalanced force. This force is provided by the action of the spine erectors. These muscles are situated near the axis of rotation (which is located near the region of the nucleus

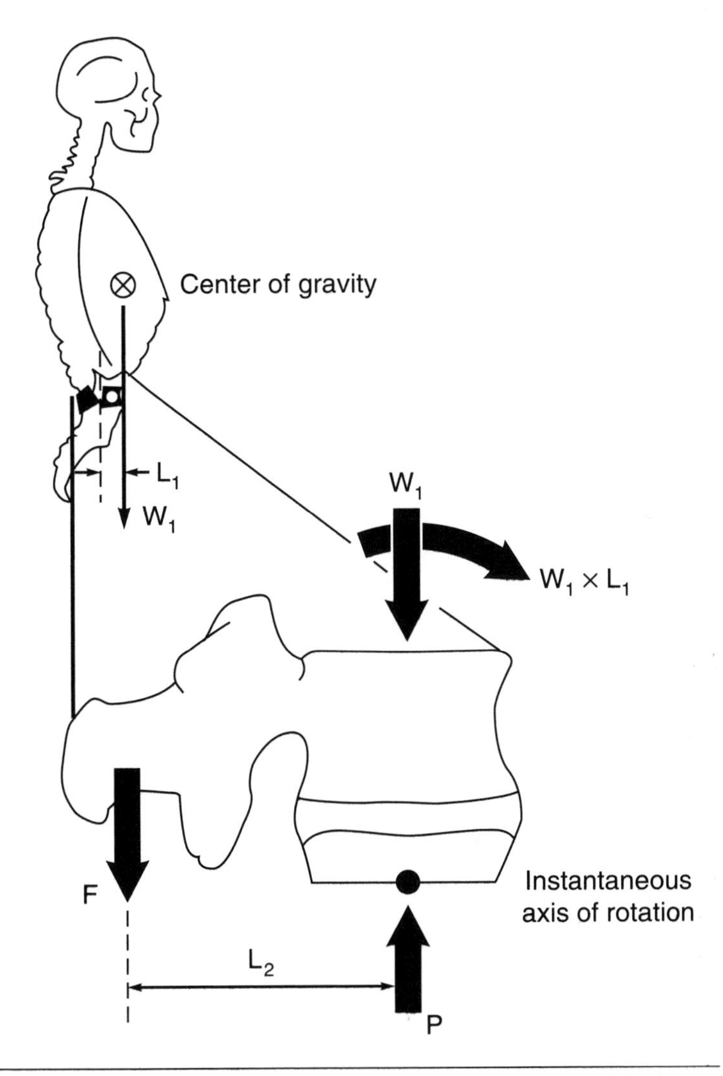

Figure 7.4 Mechanism for creating a mechanical load on the intervertebral disks. W_1, weight of the above-lying parts of the body; L_1, the lever arm; $W_1 \cdot L_1$, the flexion bending moment due to gravity; F, force of the extensor muscles of the spinal column; L_2, their lever arm. Since the system is in equilibrium, $W \cdot L_1 = F \cdot L_2$. Therefore $F = (W_1 \cdot L_1)/L_2$. The force acting on the intervertebral disc (P) is equal to the sum of the weight of the above-lying parts of the body and the muscle-pulling force, $P = W_1 + F$ or $P = W_1 \cdot [(1 + L_1)/L_2]$. Note. Adapted from *Clinical Biomechanics of Spine* (p. 50) by A. White and M.M. Panjabi, 1990 (2nd ed.), Philadelphia: J.B. Lippincott. Copyright 1990 by Augustus A. White and Manohar M. Panjabi. Adapted by permission.

pulposus of the intervertebral disk), and therefore the moment arm of the pull is slight. To produce the necessary moment of force, these muscles in turn generate considerable force (in accordance with the lever principle—the smaller the distance, the greater the force). Since the line of action of the muscle force runs almost parallel to the spinal column, this force, added to the force of gravity, sharply increases the pressure on the intervertebral disks.

Thus, a force acting on L4 in the usual upright position amounts not to half the body weight, but to the body weight. During leans, lifts, and other specific movements, external forces create a considerable moment relative to the axis of rotation that passes through the lumbar intervertebral disks. The muscles and especially the ligaments of the spinal column are close to the axis of rotation, so the force they produce sometimes exceeds the weight of the load being lifted and that of the upper parts of the body. This force contributes significantly to the mechanical load that falls on the intervertebral disks.

Role of Intraabdominal Pressure

The mechanism and the very role of the intraabdominal pressure (IAP) load have been questioned recently by some researchers. What is presented here reflects the most commonly accepted explanation.

The formula calculations cited in the caption of Figure 7.4 show that even during a lean with an 80-kg weight, the load on the lumbar vertebrae can be greater than 1,000 kg, which exceeds the limit of their mechanical

Table 7.1 Force (Body Weight) Acting on L3 in Different Situations

Posture or movement	Force
Lying, supine position, traction 30 kg applied	0.14
Lying, supine position, legs straight	0.43
Upright standing posture	1.00
Walking	1.21
Lateral trunk lean to one side	1.35
Sitting unsupported	1.43
Isometric exercises for muscles of abdominal wall	1.57
Laughter	1.71
Inclined forward 20°	1.71
Sit-up from supine position, legs straight	2.50
Lifting a 20-kg load, back straight, knees bent	3.00
Lifting a 20-kg load from forward lean, legs straight	4.85

strength. At the same time we know that athletes can lift significantly greater weights without apparent harm. Of course, this is true in part because of the considerable strength of individual anatomical structures of the spinal column in trained persons. But the main reason is one that these calculations do not take into account—the role of the internal support that emerges as a result of elevated intraabdominal pressure (IAP) during the execution of many strength exercises (Figure 7.5)

IAP increases during muscular efforts, especially during a Valsalva maneuver. As a result of internal support, the pressure on intervertebral disks can be reduced by up to 20% on average and up to 40% in extreme cases.

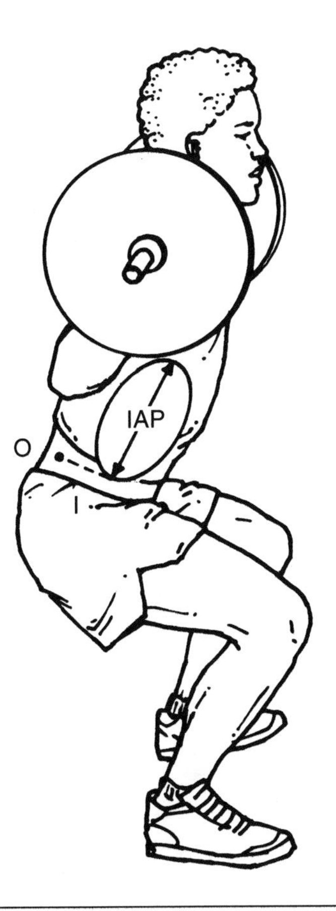

Figure 7.5 Internal support of the spinal column can be compared to the mechanical action of a ball located in the abdominal cavity. Intraabdominal pressure (IAP) produces the spinal extension moment relative to the axis of rotation O (I is the lever arm).

The most accessible method of measuring IAP is to introduce a pressure gauge into the stomach cavity. Here the intrastomachic pressure, which is almost the same as the intraabdominal pressure, is measured. Figures 7.6 and 7.7 show data on intraabdominal pressure measured during the execution of various physical exercises. On the basis of the results of several investigations, a couple of conclusions can be drawn.

We find that the IAP is proportional to the moment of force relative to the axis of rotation passing through the intervertebral disks (but not to the force produced or to the weight lifted). Because different techniques can be employed to perform identical exercises with the same weight, the externally generated force corresponds to different moments of force. Depending on the moment arm, some technique variations are more dangerous than others. We can also conclude that with an increase in the ability

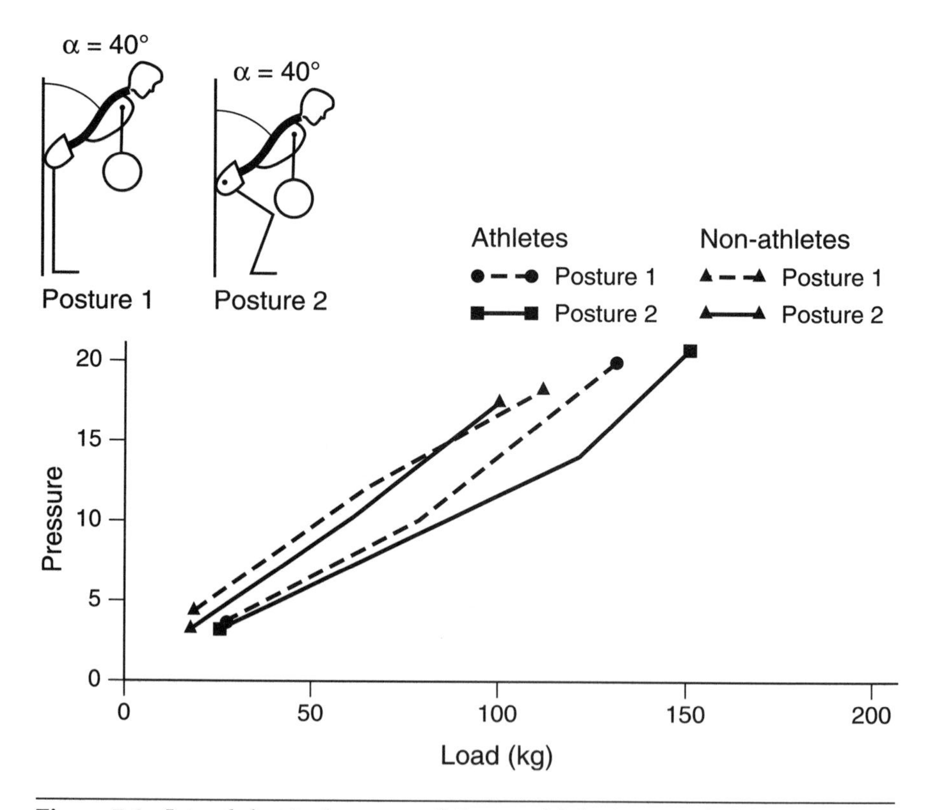

Figure 7.6 Intraabdominal pressure (kilopascals) during weight lifting. Note. From "Biomechanical Foundations in the Prevention of Injuries to the Spinal Lumbar Region During Physical Exercise Training" by V.M. Zatsiorsky and V.P. Sazonov, 1985, *Theory and Practice of Physical Culture*, (7), pp. 33-40. Reprinted by permission from the journal.

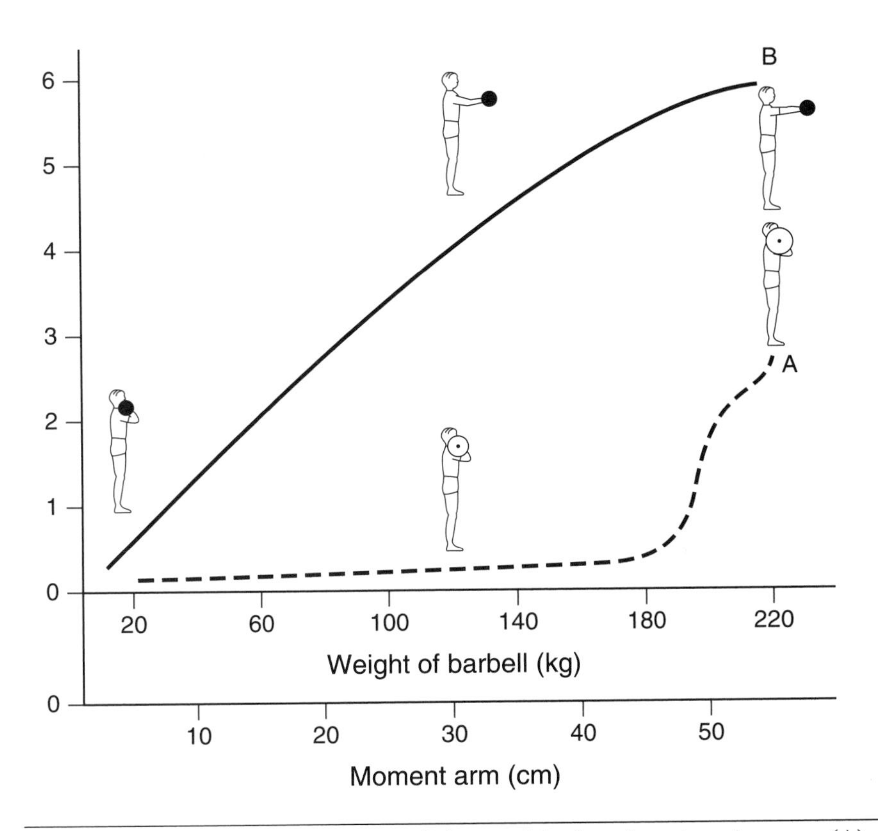

Figure 7.7 Intraabdominal pressure (kilopascals) when there is an increase: (A) in the weight borne on the shoulders; (B) in the moment arm of force (constant weight 20 kg). Note. From *Biomechanical Studies in the Prevention of Injuries to the Spinal Lumbar Region During Physical Exercise Training* by V.P. Sazonov, 1985, unpublished doctoral dissertation, Moscow: Central Institute of Physical Culture.

to lift maximal weights, IAP increases, promoting a decrease in mechanical loads affecting the spinal column.

High intraabdominal pressure is generated by the activity of muscles in the abdominal wall, the intercostal muscles, and the diaphragm. If the magnitudes of IAP and other variables (weight lifted, body posture) are measured, the amount of mechanical pressure acting on the intervertebral disks can be estimated with acceptable accuracy by specially developed biomechanical models.

Injury Prevention to the Lumbar Region

To prevent injuries to the lumbar region of the spine or reduce the consequences of these injuries, it is necessary to maximally reduce the load falling on the lumbar section of the spine and to strengthen the muscles of the lumbar region (create a "muscular corset"). People differ markedly both in the extent of muscle development in the lumbar region and in the size of the maximal loads that they can bear. Therefore, in practice, preventive advice should be strictly individualized.

From a practical standpoint, there are several important guidelines for the prophylaxis of low-back problems for athletes. It is helpful for prevention to strengthen certain muscle groups and to use proper sport technique. Some athletes may benefit from

- the use of special implements designed to decrease spinal load,
- posture correction and improvement of flexibility, and
- the use of rehabilitation procedures.

Muscle Strengthening

LBPS occurs more frequently in persons with weak or nonproportionally developed muscles such as a weak abdominal wall. Proper muscle development is required for the prevention of LBPS. In addition to strengthening erector spine muscles, athletes should exercise the muscles of the abdominal wall (not only the rectus abdominis but also the oblique muscles of the abdomen) and the short deep muscles of the back. This issue becomes complicated because it is precisely the exercises aimed at forming a "muscular corset" that are often associated with large loads on the lumbar spine. To prevent spinal overloading during strengthening of the spine erector muscles, extreme caution is necessary. This is especially true for teenagers and women. The three-year rule mentioned earlier is useful here.

Exercises for Muscles of the Abdominal Wall

Let's first analyze the load imposed on intervertebral disks in a lying position. For a person lying supine with legs outstretched, the load falling on the intervertebral disks is rather significant and is equal to approximately 35% to 40% of body weight. This is related primarily to activity of the iliopsoas (the compound iliacus and psoas magnus muscles; Figure 7.8), which apparently is manifested externally in the preservation of lumbar lordosis.

When the legs bend at the knees, the hip flexors shorten and the force of their pull drops to zero. As a result, the pressure in the intervertebral disks decreases. Pain usually disappears in patients in this position. A coach can judge the cessation of lumbar muscle activity by the disappearance of lumbar lordosis; in other words, the back becomes flat.

Exercises for the Rectus Abdominis Muscle. Muscles of the abdominal wall deserve special attention during heavy resistance training, especially

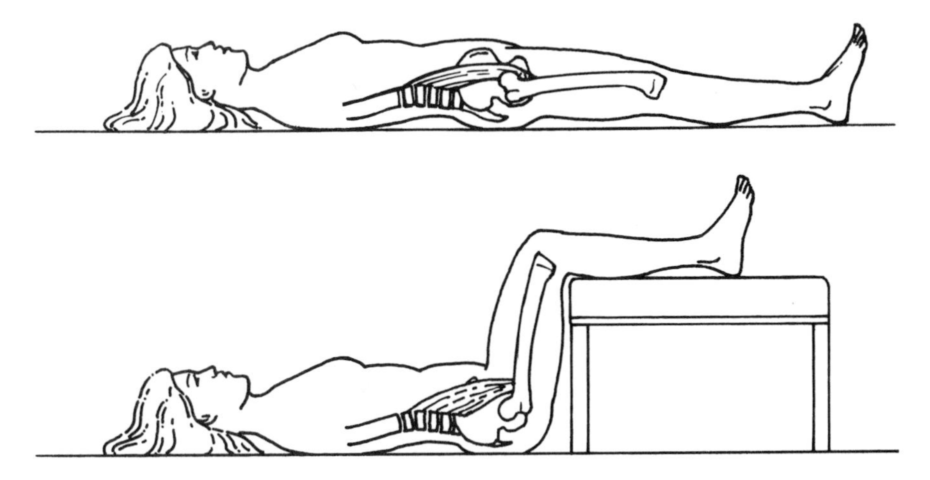

Figure 7.8 Influence of the ileolumbar muscles on the creation of pressure in the intervertebral disks. In the top panel the iliopsoas muscles are stretched; the force of their pull is applied to the spine. There is definite pressure on the disks, and lumbar lordosis is retained, due to which there is some protrusion of the edges of the disks posteriorly (see Figure 7.3 above). In LBPS patients in an exacerbation period, this position can be painful. In the lower panel the iliopsoas muscles are shortened and do not show the force of the pull. As a result, the pressure in the disks is lower; the spine straightens out in the lumbar region; and the disks do not stick out past the edge of the vertebrae. Pain usually disappears. Note. From "Biomechanical Foundations in the Prevention of Injuries to the Spinal Lumbar Region During Physical Exercise Training" by V.M. Zatsiorsky and V.P. Sazonov, 1985, *Theory and Practice of Physical Culture*, (7), pp. 33-40. Reprinted by permission from the journal.

with beginning athletes and teenagers, for three main reasons. First, the muscles that stabilize the trunk participate in locomotion as well as in many other movements. In addition, well-developed muscles of the abdominal wall help maintain proper function of the internal organs in the abdominal region. Finally, adequate strength of this muscle group is the best guarantee against an abdominal hernia (i.e., the protrusion of an internal organ or its part through the abdominal wall). A hernia may be provoked by increased intraabdominal pressure resulting from lifting heavy loads. That is, if an athlete's spine extensors are strong and the abdominal muscles are relatively weak, high intraabdominal pressure may lead to a hernia. Hernias in young athletes should be regarded as a coach's blunder. They occur when the training of abdominal muscles has been neglected.

Exercises for abdominal wall muscles fall into two groups: (a) leg raising with the torso securely anchored and (b) sit-ups, that is, raising the torso with the legs securely anchored. Leg raising in the supine position is accomplished by the activity of the flexor muscles in the hip joints (the iliopsoas muscles, the rectus femoris muscles, and others). The rectus abdominis muscle, fastened at its lower end to the pubis symphysis, is relatively inactive; it secures the pelvis and increases intraabdominal pressure. It begins to shorten only when the legs are raised high enough. At this point, however, the moment of force of gravity, pulling the legs down, is relatively slight. Since the initial pressure on the disks is rather high and the activity of the abdominal wall muscles is not significant (though it is precisely for their development that this exercise is done), this exercise is not especially valuable. Certainly it should not be the only exercise used to train the abdominal muscles.

Leg raising in a hanging position is much more effective (here the rectus abdominis muscle contracts when the moment of gravity of the legs reaches its maximum), but it is feasible only for trained persons. The so-called "basket hang" is an example of an exercise from this group. Here the performer is suspended from a horizontal bar with legs extended. The knees are drawn up to the chest until the pelvis tilts up and back, and then the trainee uncurls to the extended position.

Sit-ups are considered a major exercise for the rectus abdominis muscles. Persons at high risk of LBPS should perform sit-ups with the legs bent, as in this position the load on the spine is lighter and the effect on the abdominal wall muscles is greater. This is so because the iliopsoas muscles are in a shortened state and do not take part in generating a rotational moment of force. In sit-ups done from a straight-leg starting position, the main portion of the torque is produced by the iliopsoas muscles (which is not appropriate to the training goals here), and pressure on the intervertebral disks is very great (corresponding approximately to that for a forward lean in the upright position with a 20-kg weight in the hands). This type of exercise is hardly ever recommended for persons who have recently recovered from an attack of low-back pain.

Sit-ups should be performed with the torso in a bent position. The first step is to move the head and shoulders (thrusting the chest and abdomen forward reduces activity of the abdominal wall muscles). Note that sit-ups do have drawbacks. The abdominals are prime movers for only the first 30° to 45° of flexion movement while the hip flexors are responsible for the last 45°. Because the hip flexors are exercised through a short arc, this can induce their adaptive shortening and, in turn, hyperlordosis. Persons with LBPS can limit themselves to the first part of this exercise until the shoulder girdle becomes slightly elevated. In partial sit-ups such as this (also called partial curls or crunches), the knees are flexed to a much more oblique angle (140°-150°) and the trainee raises the trunk off the floor about 30°.

One of the exercises most frequently recommended for persons at high risk of LBPS is raising the pelvis and legs from the supine position. This exercise resembles the first part of an elbow (shoulder) stand—the "birch tree" (Figure 7.9, exercise 5). Here pressure on the intervertebral disks is small, and involvement of the abdominal wall muscles is significant.

For people suffering from LBPS and possessing a low level of muscular strength, isometric exercises are recommended. These individuals are advised to begin training of the "muscular corset" after an aggravation of LBPS. The value of these exercises is that they put a certain load on the muscles of the abdominal wall with almost no increase in pressure on the intervertebral disks. To do the exercises, after a normal inhalation the person contracts the musculature of the abdominal wall and back with the glottis closed and the rectal sphincter contracted, trying to produce a strong exhalation. Since this kind of straining is created through the action of the musculature of the trunk and diaphragm, multiple repetitions elicit a training effect. The exercise should be repeated 10 to 15 times with the muscle contraction lasting 3 to 5 s. This series should be repeated three to four times a day. In experiments conducted with a double-blind control, isometric exercises have been shown to produce a decidedly better effect than other types of exercises.

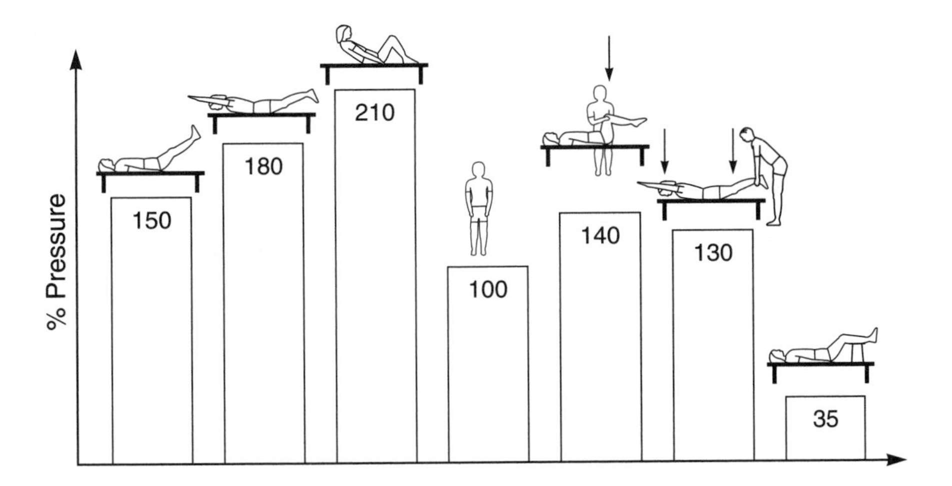

Figure 7.9 Intradisk pressure (in % of pressure relative to the upright posture) in several exercises for strengthening the "muscular corset." Note. From "The Lumbar Spine, an Orthopaedic Challenge" by A.L. Nachemson, 1976, *Spine*, **1**, pp. 59-71.

Exercises for the Oblique and Internal Abdominis Muscles. In many movements, such as symmetrical weight lifts, high IAP is created as a result mainly of activity of the oblique and internal abdominis muscles rather than the rectus abdominis. The reason is that the rectus abdominis, while active, generates a trunk-bending moment that should be counterbalanced by an additional moment produced by the co-contracted muscles—the spine extensors. The higher the activity of the rectus femoris muscles, the greater the IAP (which is good). At the same time, however, the higher the activity of these muscles, the greater is the bending moment that must be overcome by the spine extensors in order to produce the required extensor moment of the spine. As a result, the trunk flexors are modestly activated during weight-lifting tasks. The IAP is generated chiefly by activity of the oblique abdominis muscles (and diaphragm). In addition, strong oblique muscles reinforce the erector spinae fasciae. The fasciae support the spine and reduce strain on the back extensor muscles. So exercises for the oblique and internal abdominis muscles, such as trunk rotations against resistance and lateral sit-ups (trunk lifts), should be included in training protocols.

Exercise for the Short, Deep Muscles of the Back

Muscles of the lumbar region (specifically, the epaxial muscles, such as the interspinales connecting adjacent spinous processes or the intertransversalis connecting adjacent transverse processes of the vertebrae) are difficult to activate in ordinary physical exercises.

The following exercise is recommended for training these muscles. The athlete stands with the back against a wall so that heels, buttocks, shoulders, and the back of the head touch the wall. The next step is to address the lumbar lordosis by completely straightening the spine so that the lumbar region rests against the wall and even exerts pressure on it. Here, contact between the wall and the other parts of the body should continue (Figure 7.10). This exercise often proves difficult for even highly skilled athletes. When this is the case, it can be tried in a supine position. After mastery, it can be done without the aid of a wall. The usual pattern is five to six attempts of 4 to 5 s each.

Requirements for Proper Sport Technique

When the body is inclined forward, the activity of muscles that extend the spinal column increases at first; then, with a deeper lean, this activity almost completely disappears (see Figure 7.11). The ligaments and fasciae of the back assume the load here. Since they are close to the axis of rotation, they should generate considerable force to counteract the force of gravity moment. Here, pressure on the intervertebral disks is very high.

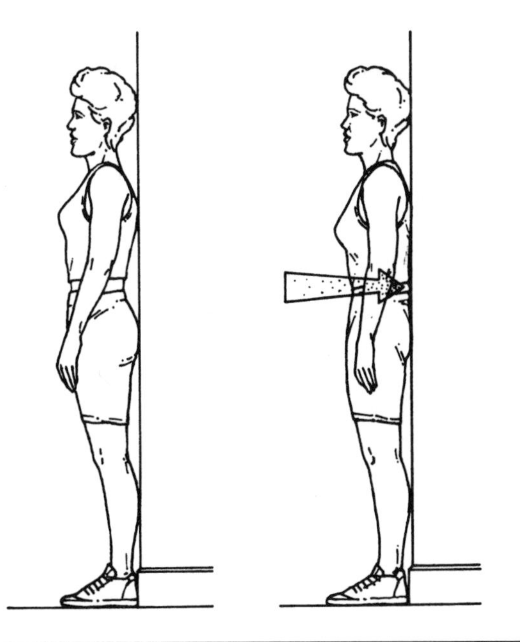

Figure 7.10 An exercise for the short, deep muscles of the back (so-called "pelvic tilt").

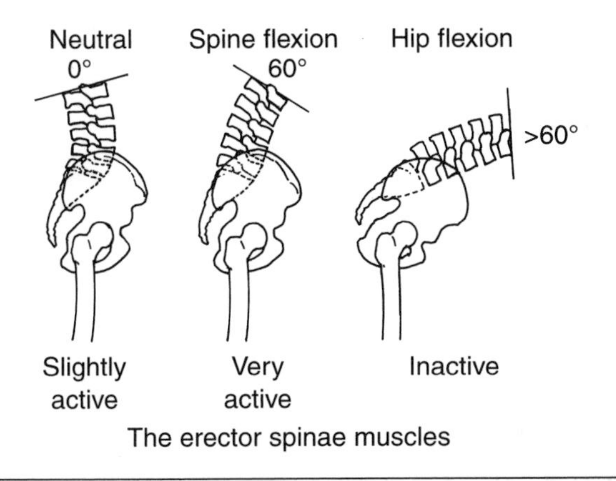

Figure 7.11 The activity of muscles that support the spinal column during the execution of a forward lean. Lumbar flexion accounts for approximately 45° to 60° of motion until the posterior ligaments become taut. The second part of the movement is performed due to the pelvis rotation until the pelvis is passively restricted by the gluteus and hamstrings. In this position, no muscular activity is seen. The trunk weight is counterbalanced by passive forces of the erector spinae fasciae, posterior ligaments, and muscles. Note. From "Biomechanical Foundations in the Prevention of Injuries to the Spinal Lumbar Region During Physical Exercise Training" by V.M. Zatsiorsky and V.P. Sazonov, 1985, *Theory and Practice of Physical Culture*, (7), pp. 33-40. Reprinted by permission from the journal.

A "rounded back" position is dangerous in lifting weights because, as a result of lumbar spine flexion, the compression load acts on the anterior part of the intervertebral disks while the extension load acts on the posterior part. Specifically, a pressure concentration takes place. This pressure, that is, the amount of force falling on a unit of the disk surface, is very considerable (Figure 7.12).

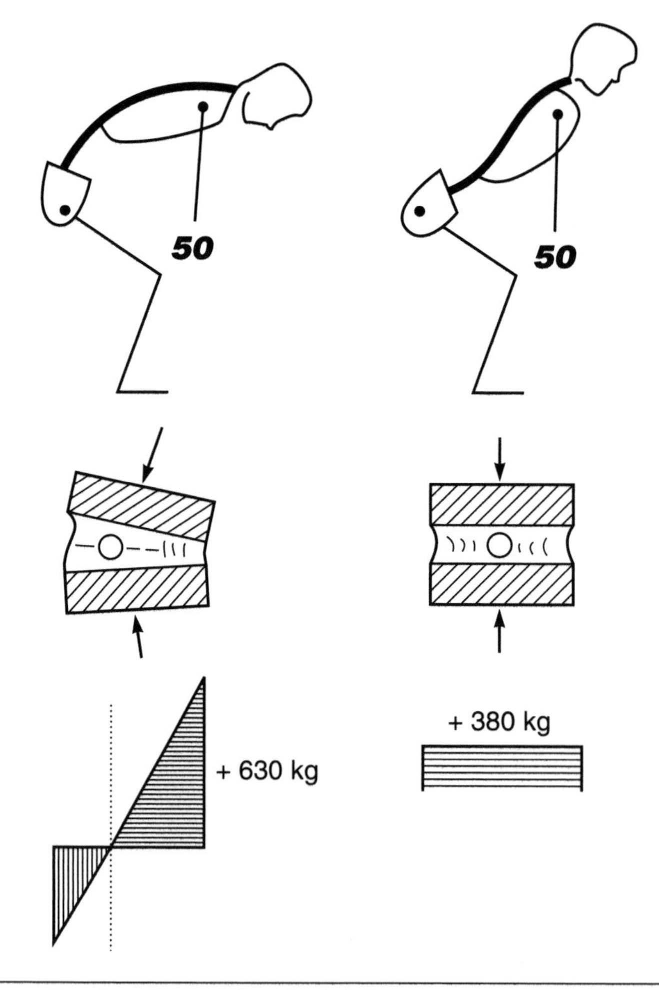

Figure 7.12 Load on the intervertebral disks when 50 kg is lifted by different methods. Left, incorrect technique ("rounded back"); right, correct. Compression loads on a lumbar intervertebral disk amount to 630 kg and 380 kg, respectively. Note. From *Motor Abilities of Athletes* (p. 60) by V.M. Zatsiorsky, 1966, Moscow: FiS. Reprinted by permission from Fizkultura i Sport.

Some practical advice is, first, to preserve lumbar lordosis when lifting weights (Figure 7.13). In addition, if possible, lift weights while squatting rather than stooping. This is a principle that should be learned from childhood so that proper methods of lifting weights become a habit. Physical education should stress developing the extensor muscles of the legs so that subjectively it is as easy for a person to squat as to stoop.

Teenage athletes (in sports such as tennis, basketball, and volleyball) often neglect conditioning training, including strength development, during the initial stages of multiyear preparation. At 20, they find that their athletic performance is limited by poor physical fitness. Then they try to develop strength as fast as possible by copying the training patterns of athletes from other sports such as track and field, especially with free weights. But track and field athletes at the age of 20 have had several

Figure 7.13 Correct (left) and incorrect techniques of leaning and weight lifting. Note. From "Biomechanical Foundations in the Prevention of Injuries to the Spinal Lumbar Region During Physical Exercise Training" by V.M. Zatsiorsky and V.P. Sazonov, 1985, *Theory and Practice of Physical Culture,* (7), pp. 33-40. Reprinted by permission from the journal.

years of experience in conditioning. It is impossible for novices to replicate their training routine. It is simply dangerous.

An unfortunate example concerns a conditioning coach who was invited to work with the U.S.S.R. women's tennis team in the early 1980s. He had never worked with comparable athletes before and had no conception of their inexperience with strength training. The training routine he recommended duplicated heavy resistance programs from other sports. The result? In 6 months, 9 of 10 athletes had low-back problems. Eight of them never rehabilitated completely and dropped out of international sport.

■ *Be Aware Checklist*

Extreme caution is in order when weight lifting is executed by women, very tall men, and teenagers. Check:

- Do these athletes have an immediate need for weight training? Why free weights? Strength exercises without free weights are innumerable. Be creative.
- Do your trainees have proper prior experience in strength training (without a barbell)? Recall the three-year rule. Is this principle satisfied?
- Strengthen trunk muscles—spine erectors and abdominal muscles.
- Use exercise machines first, and then free weights.
- Teach correct lifting technique. Monitor the lifting pattern.
- Begin with small loads. Inappropriate weight, rather than the barbell itself, is the source of risk. For a majority of inexperienced athletes, a bar without added plates provides adequate resistance.
- Use weight-lifting belts and bolsters.
- Teach proper breathing patterns.

Good luck! Be cautious.

Implements

Several types of implements can be used to enhance IAP and fix the lumbar spine. One of these is a special bolster for use in the performance of exercises for the muscles that extend up the spinal column (Figure 7.14). Weight-lifting belts are also recommended to increase the IAP and reduce the load on the spine. By tradition, weight-lifting belts are constructed with the idea of providing support against spinal deformation. This is important in an exercise such as a standing barbell press. This exercise, however, has been excluded from the Olympic weight-lifting program.

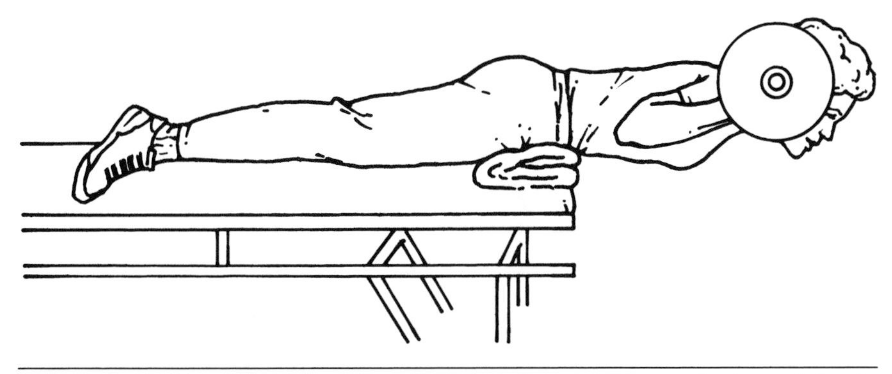

Figure 7.14 Use of a pad placed under the abdomen increases the intraabdominal pressure (IAP) and lessens the load on the intervertebral disks. Note. From "Biomechanical Foundations in the Prevention of Injuries to the Spinal Lumbar Region During Physical Exercise Training" by V.M. Zatsiorsky and V.P. Sazonov, 1985, *Theory and Practice of Physical Culture*, (7), pp. 33-40. Reprinted by permission from the journal.

Even so, the construction of waist belts for weight lifting remains the same. According to some research, belts that support the abdomen, rather than the spine, increase the IAP and consequently decrease spinal load to a greater extent (Figure 7.15).

Posture Correction and Flexibility Development

Increased lumbar lordosis gives rise to a higher risk of LBPS. Lordosis compensates the obliquity of the sacrum, which is tilted with respect to the vertical. The position of the sacrum is characterized by a sacrovertebral angle formed by the upper surface of the first sacral vertebra and the horizontal. Normally, the smaller this angle, the better. A more vertical position of the sacrum favors stability at the lumbosacral junction.

The slant of the sacrum can be corrected by the proper strength development of corresponding muscles. (Note that in heavier people the sacrum is usually directed more obliquely because of the weight of the body bearing on it, and in this case the first recommendation is to lose weight.) The corresponding muscles are

- trunk flexors (rectus abdominus) and hip extensors (hamstring)—these muscles, when activated, tend to decrease the sacral angle, rotating the sacrum in a more vertical position; and
- trunk extensors and hip flexors (rectus femoris)—these rotate the trunk into a more horizontal position.

In athletes who perform many barbell squats and sit-and-reach exercises but neglect to strengthen the abdomen and stretch the hip flexors,

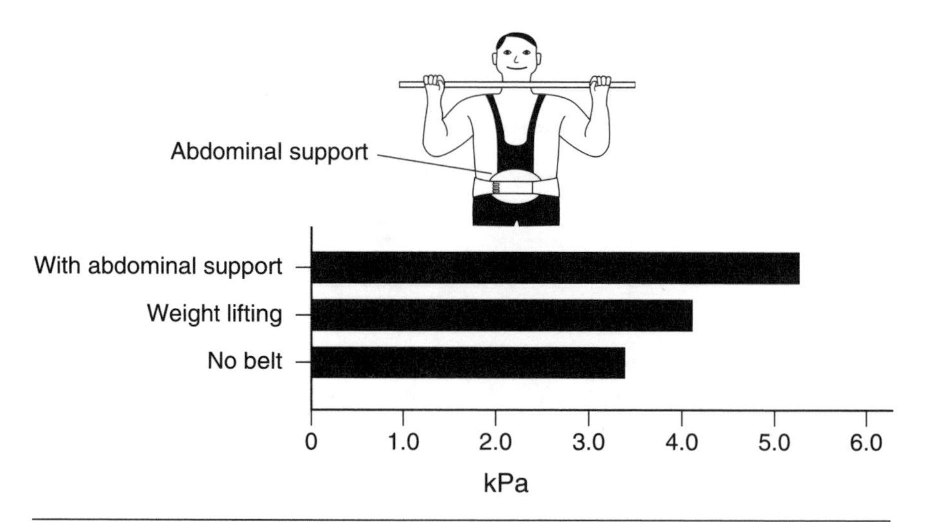

Figure 7.15 Intraabdominal pressure (IAP) during exercise under different conditions. Top, a (patented) belt with firm abdominal support (Russian patent #1378834 to V.M. Zatsiorsky and V.P. Sazonov from November 8, 1987). Bottom, IAP while lifting two 10-kg dumbbells (shoulder flexion with the arms stretched) under three conditions: belt with abdominal support, ordinary weight belt, and no belt. Note. From "Belt-Corsets Reducing Risk of the Spine Lumbar Trauma at Weight Lifting and Strength Exercises" by V.M. Zatsiorsky and V.P. Sazonov, 1987, *Theory and Practice of Physical Culture*, (3), pp. 15-18. Reprinted by permission from the journal.

the hamstrings are often flexible while the abdominal muscles are weak and the hip flexors are tight. In this case the anterior pelvic tilt becomes exaggerated. In turn, compensating hyperlordosis appears, causing the disks to bulge posteriorly and putting compressive stress on the vertebral facets. The nerve roots that exit from the vertebrae can be compressed and this can lead to pain. To correct pelvic tilt and hyperlordosis, the advice is to strengthen the abdominal muscles and perform stretching exercises to decrease tightness of the hip flexors.

■ *Have Lower Back Problems?*

First, consult a physician. Ask for a diagnosis. Usually an X-ray, nuclear magnetic resonance imaging (NMRI), or both, are required. Keep the results for future reference. If nothing serious is discovered and training is permitted, then take these steps:

Step 1—For an acute pain period. Perform relaxation exercises for at least 1 to 2 weeks and do the isometric exercises described in the section on exercises for the rectus abdominis muscle. The goal of relaxation exercises is first to decrease and then to completely eliminate muscle spasm. Here are examples from a relaxation routine:

- Lie down. Relax facial muscles. Relax eyelids. The eyes should be semiclosed during the entire routine.
- Relax neck muscles. Permit your head to fall down freely to the right without any muscular resistance. Only gravity is acting. Wait 3 s; rotate your head face up. Relax again. Permit the head to fall down to the left. Repeat three to four times on each side.
- Bend the right knee with foot on the floor. Relax. Permit the leg to extend, the foot gliding along the floor. Again, only gravity force is involved. Repeat with each leg three to five times.
- Bend an arm. Relax. Permit the arm to fall down. Repeat with the second arm. Relax. Repeat several times.
- Perform isometric exercises for the abdominal muscles.
- Repeat the relaxation routine in reverse order. Relax, relax, relax.

Step 2. When the pain disappears. Temporarily decrease the load on the lumbar spine (e.g., use leg lifts instead of squats). Then analyze these factors:

• Your training routine	Did you overload the spinal region? Did you squat much the last time? Did you perform many dead lifts?
• Your fitness level	(a) Are your spine erectors, rectus abdominis, oblique abdominis, and epaxial muscles strong enough? Did you neglect to strengthen them? (b) How is your flexibility? Can you touch the floor? with your palms? Are your hip flexors tight? (c) Is your pelvis inclined much in your customary posture? Do you have large lumbar lordosis?
• Your lifting technique	Is your spine rounded during lifting? Ask somebody to check it.
• Your abdominal support	Do you wear a waist belt when lifting? Does this belt have an abdominal support? Do you use bolsters?

- Your restoration measures — What kind of restoration measures do you usually use between training workouts? None? This is not advisable.

Depending on the answers, prescribe corrective and preventive measures for yourself. Reread this chapter carefully and decide what suits you best. Follow the new routine. When these measures are taken, nine of ten athletes completely restore their abilities and experience no difficulty or have only minor problems with their spines.

Rehabilitation Procedures

To restore the dimensions and properties of compressed intervertebral disks created by exposure to large systematic loads (weight lifting, rowing), restorative measures are usually recommended. These include massage and swimming in warm water. When the load falling on the intervertebral disks is reduced, the degree of disk hydration increases (Figure 7.16).

Many coaches recommend alternating weight lifting with hangs during a training session. However, spine length in the majority of athletes decreases during such hangs. Usually a reflex activation of the trunk muscles takes place and these contracted muscles prevent the spine from lengthening. Consequently, the dimensions of the intervertebral disks are not restored. Not all athletes can relax in the hanging position. In addition, full disk hydration occurs only during prolonged removal of the compressive load acting on the spine, and this does not happen with hangs alone.

Spinal traction has proven to be a much more effective procedure. Figure 7.17 shows the recommended posture and unit for this stretching. Spinal traction, performed twice a week with individually adjusted traction force (up to 100 kg for elite weight lifters from the super heavy weight class), is a very useful restorative measure.

Spinal traction is recommended only to athletes with no history of LBPS. Preliminary medical investigation and permission from a physician are required. When an athlete is already suffering from LBPS, traction can be a negative influence. Figure 7.18 shows a reason for this. During traction, the lumbar lordosis diminishes and the spinal column takes on a straighter position. Here there is a relative shift of the spinal cord rootlets in a caudal direction. Therefore, if disk protrusion has occurred above a rootlet, traction alleviates the pain (right); but if it is under the rootlet, the pain is exacerbated (left). A physician needs to make the decision whether spinal traction is advisable for a given athlete.

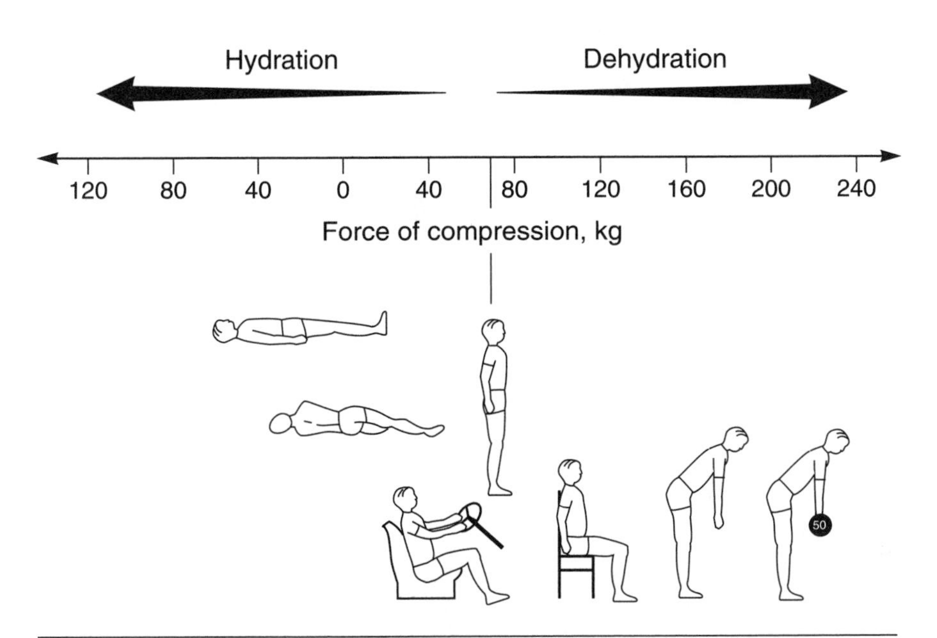

Figure 7.16 Intradisc pressure and water saturation of nucleus pulposus (for L3 disk). Note. From "Towards a Better Understanding of Back Pain: A Review of the Mechanics of the Lumbar Disc" by A. Nachemson, 1975, *Rheumatology and Rehabilitation*, **14**, pp. 129-143.

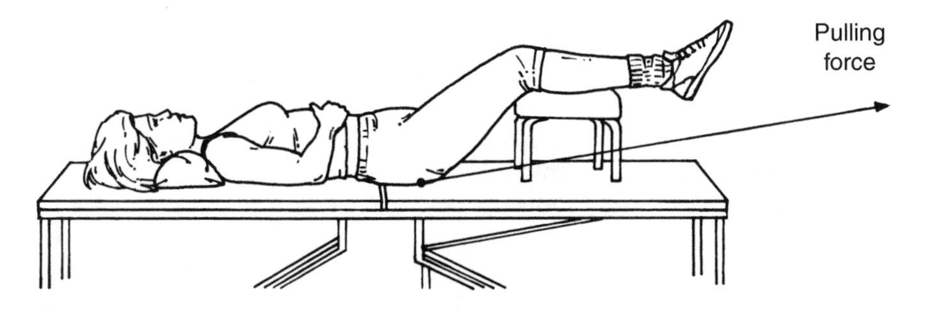

Figure 7.17 An apparatus (called a split table) used for spinal traction. The athlete's legs are bent, and the force of the pull is oriented at an angle to the horizontal (in order to keep the back flat). Note. From "Spinal Traction as a Rehabilitation Tool" by V.M. Zatsiorsky and S.S. Arutiunjan, 1987, Scientific Information. Moscow: Central Institute of Physical Culture.

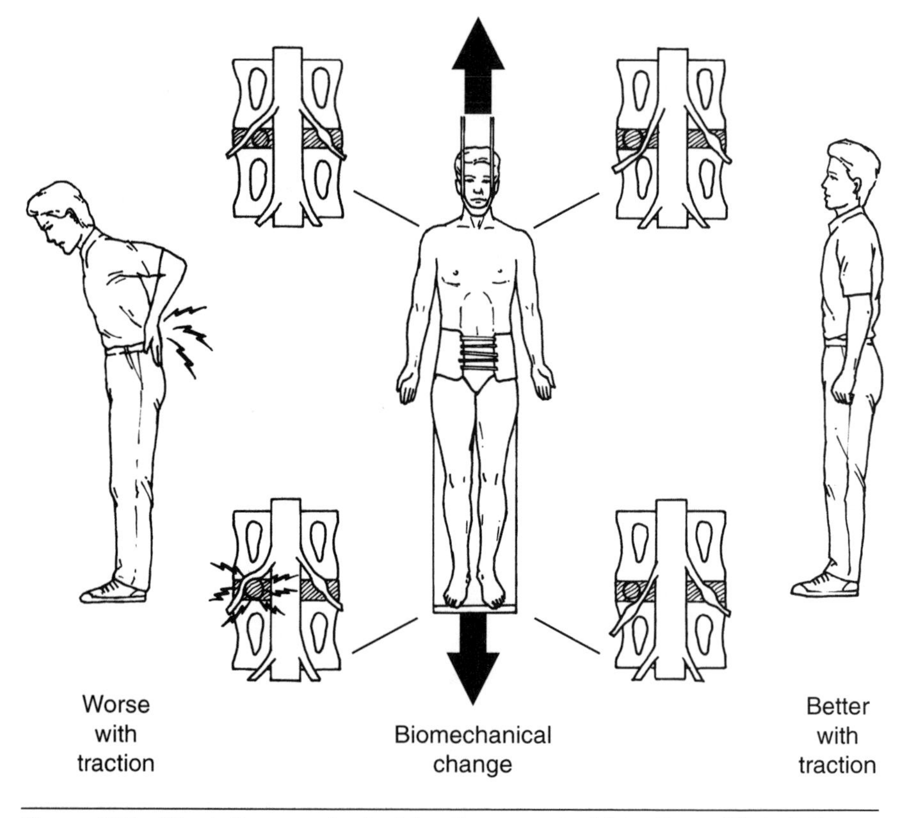

Figure 7.18 The influence of spinal fraction on pain. Note. From *Clinical Biomechanics of the Spine* (p. 434) by A. White and M.M. Panjabi, 1990 (2nd ed.), Philadelphia: J.B. Lippincott. Copyright 1990 by Augustus A. White and Manohar M. Panjabi. Reprinted by permission.

Summary

Coaches and athletes should give special attention in heavy resistance training to prevention of injury to the lumbar spine region.

Biomechanically, intervertebral disks are characterized in large part by water saturation and intradiskal pressure. During a vertical load the mechanical strength of disks is adequate, not inferior, to the strength of adjoining vertebrae. But during a lean of the spinal column, the nucleus pulposus is shifted to the side opposite the lean and the fibrous ring is somewhat protruded. This can induce compression of the spinal cord rootlets and give rise to a painful sensation.

Mechanical loads affecting the intervertebral disks are classified as impact and static. Impact loads are typically experienced during landing.

An impact load during landing is softened by the combined influence of the properties of the supporting surface, the quality of the footwear, the dampening properties of the motor system, and the landing techniques. "Soft" landing techniques, in which ankle plantar flexion and knee flexion are coordinated, reduce the magnitude of impact forces.

Static loads acting on intervertebral disks are mainly generated by muscle tension and tendon forces rather than by the external load itself. During weight lifting, an extremely high load on the lumbar vertebrae can be reduced somewhat by elevated IAP, which acts as an internal support. As a result, the pressure on intervertebral disks can be reduced nearly 20% on average and up to 40% in extreme cases. When the ability to lift maximal weights increases, IAP increases, promoting a decrease in mechanical loads affecting the spinal column.

To prevent injuries to the lumbar region of the spine, it is necessary to maximally reduce the load falling on the lumbar region and strengthen the muscles of the region (create a "muscular corset"). (Among the prophylactic measures are muscle strengthening and proper sport technique.)

Muscle groups that need to be strengthened, in addition to erector spine muscles, are the abdominal wall muscles and the short deep muscles of the back. Proper sport technique also prevents injury. Lumbar lordosis should be preserved when weights are lifted. If possible, weights should be lifted while squatting rather than stooping.

Implements can enhance IAP and fix the lumbar spine. Two of these are waist belts, especially those with a firm abdominal support, and bolsters. Posture correction and flexibility development are also recommended, especially for people with increased lumbar lordosis. To correct pelvic tilt and hyperlordosis, the abdominal muscles must be strengthened and tightness of the hip flexors must be decreased.

To restore the dimensions and properties of intervertebral disks compressed by exposure to large, systematic loads, some rehabilitation measures are useful. These include massage, swimming in warm water, and, especially, spinal traction.

CHAPTER 8

Goal-Specific Strength Training

Heavy resistance training is used for different purposes. It has applications to specific goals, including strength performance, power performance, muscle mass, endurance, and injury prevention. In this chapter we examine the peculiarities of various strength training methods. Usually heavy resistance is used to enhance muscular strength—that is, the maximal force that can be generated by a trainee in a given motion. For example, in Olympic sports weight lifting meets this criteria. We have already looked at training for strength purposes, but before discussing other purposes, it will be useful to make some comments about the training of experienced strength athletes.

Strength Performance

The general idea with training experienced athletes is not to train strength itself as a unified whole; rather, it is to train the underlying factors, both muscular and neural. To improve neuromuscular coordination (motor unit recruitment, rate coding, synchronization, the entire coordination pattern), the maximal effort method is the first choice. On the other hand, to stimulate muscle hypertrophy, the methods of repeated and submaximal efforts are more appropriate. By varying the type of exercise, its intensity (method of training), and its training load (volume), we can induce positive adaptation in the desired direction. Conversely, standard exercises and a constant load elicit only premature accommodation and staleness.

All three facets of heavy resistance training (i.e., exercise type; training method—maximal vs. submaximal efforts; and training volume), should be changed in a concerted manner. Because the superposition of training efforts among different heavy resistance methods is not negative, these methods may in principle be combined in one microcycle and even in one training day and session. For instance, it is possible to lift a 1-RM barbell and then use the method of submaximal effort in the same workout. However, the proper timing of exercises, methods, and loads over time brings better results. In the typical timing pattern, an exercise complex is changed once every two mesocycles. For instance, only two or three snatch-related exercises, of nine total, are used during two consecutive mesocycles. The snatch-related exercises are classified according to the type of motion and the initial barbell position. The types of motion are (a) competition snatch (barbell is lifted and fixed in a deep squat position), (b) power snatch (barbell is caught overhead with only slight leg flexion), and (c) snatch pull (barbell is only pulled to the height and not fixed). There are three initial barbell positions: (a) from the floor, (b) from blocks positioned above the floor, and (c) from the hang. Thus there are in total nine combinations.

In this typical pattern of timing, the dominant methods are changed every mesocycle with the routine during the first mesocycle directed primarily at inducing muscle hypertrophy (mainly by the methods of submaximal and repeated efforts). The training load is varied, usually according to the empirical "60% rule."

■ *Training Goal: Maximal Strength*

Combine high-intensity training (to improve neuromuscular coordination) and the repeated or submaximal effort methods, or all, to stimulate muscle hypertrophy. Change the exercise batteries regularly. Vary the training load.

Power Performance

In many sports, strength exercises are performed with the main objective to improve power, or the velocity of movement, against a given resistance (body weight, implement mass) rather than maximal strength itself. In such situations, maximal strength is regarded as a prerequisite for high movement speed. However, the transmutation of acquired strength gains into velocity gains is not easy. Two issues are of primary importance: the proper selection of strength exercises and training timing.

The requirements for exercise specificity should be thoroughly satisfied. The exercise of first choice should be the main sport exercise with additional resistance (note that we are discussing the training of qualified athletes, not novices)—see chapter 6, sport exercises with added resistance. This resistance should be applied in the proper direction (in locomotion, horizontally) and not exceed a level at which the motion pattern (the sport technique) is substantially altered.

Before a training period, it is advisable to test athletes in the main sport exercises with additional resistance (as well as with decreased resistance, if possible) to determine at least part of the resistance (force)-velocity curve for each trainee (Figure 8.1). For instance, shot-putters can be tested

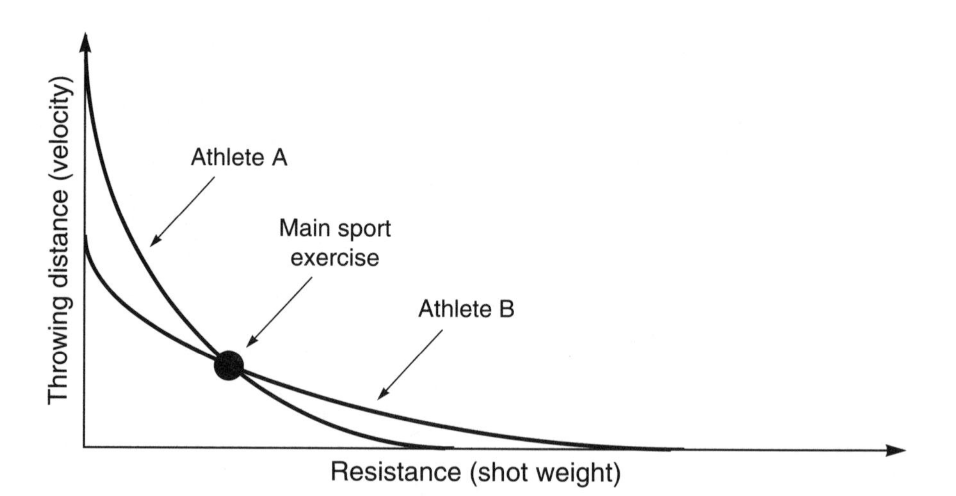

Figure 8.1 Testing results for two athletes having the same sport performance in puts of the standard shot. The athletes' achievements in throwing heavy implements are different. Since athlete A has lower results with heavy implements than athlete B, it is possible to conclude that his strength potential may be greatly improved by exercising with heavy implements. Athlete B must pay more attention to other training directions (polishing sport technique, putting light implements, and the like).

with both a standard implement (7,257 g) and with shots of 8- and 9-kg weights. Using the data on outstanding athletes as norms and comparing them with individual testing results, you can recommend the appropriate training protocol for the given period—whether the athlete should pay primary attention to training with heavy or light implements.

The coach and athlete, when selecting strength exercises for power training, should be attentive to all the facets of exercise specificity described in chapter 6 (working muscles, type of resistance, time and rate of force development, velocity of movement, direction of movement, and the force-posture relationship). Working muscles should be the same as those in the main sport exercise and the type of resistance should mimic the main sport exercise as much as possible. Specifically, it is usual to recommend isokinetic exercises characterized by a slow speed of motion and a smooth, protracted force generation for dryland training in aquatic sports (swimming, rowing, canoeing) but not for power sport disciplines. In contrast, strength exercises with free weights should be restricted in the training of swimmers since these do not permit muscle relaxation immediately after the effort.

If the time available for force development in a sport exercise is short (less than 0.3 s), the rate of force development rather than maximal strength itself is a deciding factor. Comparing maximum force production with the maximum attainable force in the fast movement has proven to be a useful tool for planning training. If the explosive strength deficit (i.e., the difference between maximum strength and the force values generated during a sport movement at takeoff, delivery phase, etc.) is too high (more than about 50% of F_{mm}), heavy resistance training directed toward the enhancement of maximum strength is not efficient; maximum strength gain is of no value toward increasing the velocity (power) of the motion. Because of the short duration of effort, maximal force values cannot be generated, so the rate of force development (RFD) rather than maximum strength has to be the primary training objective.

Maximal concentric efforts like the lifting of maximal loads can enhance the RFD in some athletes. However, because such motor tasks require maximal force rather than maximal RFD, this method may not bring positive results to highly trained athletes.

To enhance RFD, exercises with maximally fast bursts of muscle action against high loads are used. Since the load is high, movement velocity may be relatively low, but the muscle action velocity (RFD) must be extreme. The bursts of muscle action should be performed as fast as possible with maximum voluntary effort. These exercises are done in a rested state, usually immediately after a warm-up. The typical routine consists of three sets with three repetitions against a load of about 90% of maximum. Rest intervals between sets should be long (about 5 min). Other muscle groups can be exercised during the rest intervals. When the training objective is to improve RFD, these exercises are commonly performed four times a week; to retain the RFD, twice a week. Because of accommodation, after 6 to 8 weeks of such training the exercises should be changed.

The rate of force development can also be improved during the training of reversible muscle action (see discussion of the stretch-shortening cycle later in this chapter).

Movement velocity is the next important feature of strength exercises used to enhance power. The typical objective in this case is to increase the velocity of a performed movement against a given resistance. In a force-velocity diagram, this appears as a shift of the corresponding force-velocity value from point F-V_1 to point F-V_2 (Figure 8.2a). However, it is impossible to change the position of any one point on the force-velocity curve (i.e., the movement velocity with the given resistance) without altering the position of the entire curve (i.e., the velocity with different resistances). Four variations on changing the force-velocity values are possible.

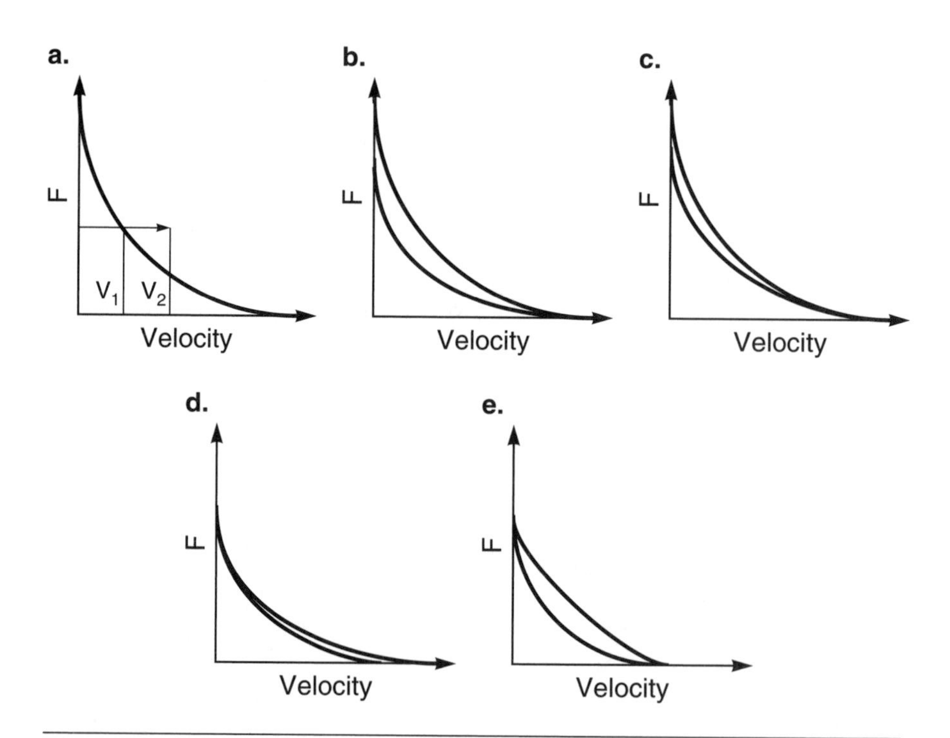

Figure 8.2 Changes of force-velocity curves resulting from training. See explanation in text.

In the first variation (Figure 8.2b), the positive velocity gain appears over the entire range of the force-velocity curve. If this were a force-velocity curve for a throwing task (in which the mass of the implement was changed and the throwing distance was measured), the force-velocity curve change means that after training, the athlete could throw farther using both heavy and light implements. This variation is typical for young athletes and is rarely seen with experienced ones. Training with exercises executed in a high-resistance, low-velocity range favors a gain in movement velocity with high resistance (Figure 8.2c), and performance results with heavy implements are mainly improved. This is the most typical way to improve athletic performance. The third variation, training with a low-resistance, high-velocity demand, brings forth improvement in the low-resistance zone (Figure 8.2d). This is a useful but auxiliary training strategy to be alternated with high-resistance training (during or immediately before a tapering period).

Finally, training in the intermediate range of resistance (e.g., with a main implement only) leads to a straightening of the force-velocity curve (Figure 8.2e). Here performance results improve in the median span of the curve. This happens as an outcome of specific training with constant implements. With this pattern of force-velocity gain, the performance is only briefly improved (usually for no more than one season), and the magnitude of gain is relatively small. The force-velocity curve can become straight but it cannot become convex. To substantially improve performance at a given resistance, achievements in the high-resistance or low-resistance zones must also be enhanced. This situation is rather controversial. On the one hand, training results depend on exercise velocity, and in order to improve the velocity with standard resistance, an athlete must exercise in the same force-velocity range as in the main sport exercise. This specific training elicits the force-velocity curve change shown in Figure 8.2e, but this change represents only a short-term effect. On the other hand, a substantial performance improvement requires less specific exercises in the high-resistance, low-velocity domain as well as in the low-resistance, high-velocity domain. These considerations are confirmed by the training practice of elite athletes (Figure 8.3).

The direction of applied movement is a key determinant of exercise effect. In many movements, muscles are forcibly stretched before shortening (stretch-shortening cycle, reversible muscle action). As described in chapter 2, the underlying mechanisms of the reversible muscle action are complex. For this reason reversible muscle action is specific (especially in highly trained athletes) and should be trained as a separate motor ability (similarly, in this respect, to anaerobic endurance and rate of force development). The exercises used for this purpose are described in chapter 6.

In conclusion, strength training for power production is composed

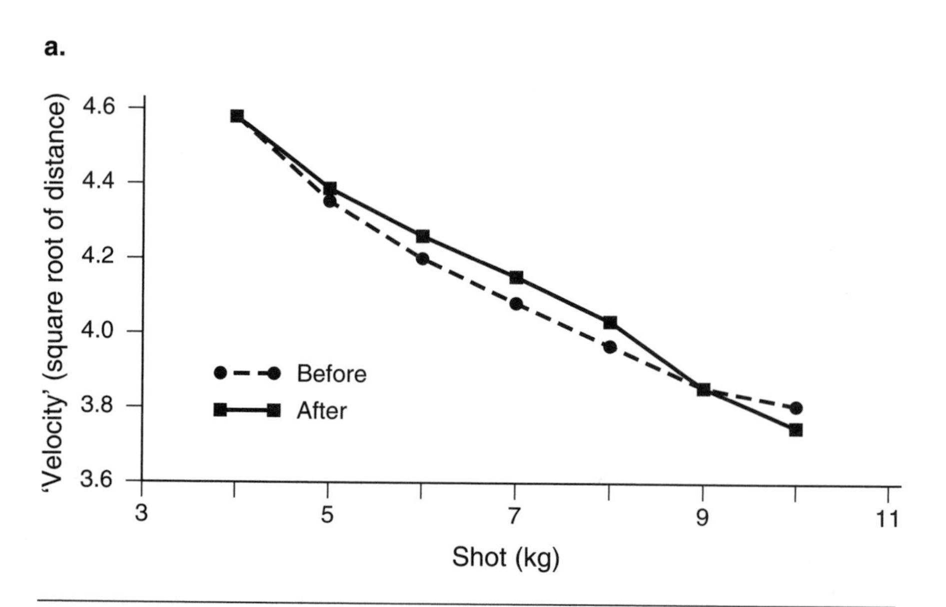

Figure 8.3 Performance results in standing shot-putting before and after 7-week training with different shots; 4- to 10-kg shots were used for testing. (a) Standard shots; only 7,257-g shots were used ($n = 4$). (b) Heavy shots (8-10 kg); a throwing routine consisting of heavy shots (70% of all the puts) and standard shots (7.257 kg, 30%) ($n = 4$). (c) Light shots (4.5-6.0 kg); the puts of light shots comprised 70% of all efforts ($n = 3$). In shot-putting, the throwing distance is the function of the release velocity (v_0), angle of release (α), and the height of release (h):

$$\text{Distance} = \frac{v^2}{g} \cdot \cos\alpha \left(\sin\alpha + \sqrt{\sin^2\alpha + \frac{2gh}{v^2}} \right),$$

where g is the acceleration due to gravity. As the distance is the quadratic function of release velocity, the square root of distance, plotted along the ordinate axis, represents (approximately) the velocity at release. Note. From *The Use of Shots of Various Weight in the Training of Elite Shot Putters* by V.M. Zatsiorsky and N.A. Karasiov, 1978, Moscow: Central Institute of Physical Culture.

of (a) main sport exercises with added resistance and (b) assistance exercises. The latter are directed toward the development of (a) maximal strength, (b) rate of force development, (c) dynamic strength (the muscular force generated at a high velocity of movement), and (d) force produced in stretch-shortening (reversible) muscle action. The proportion of exercises from these groups should be determined individually for each athlete and should change when the athlete's status changes.

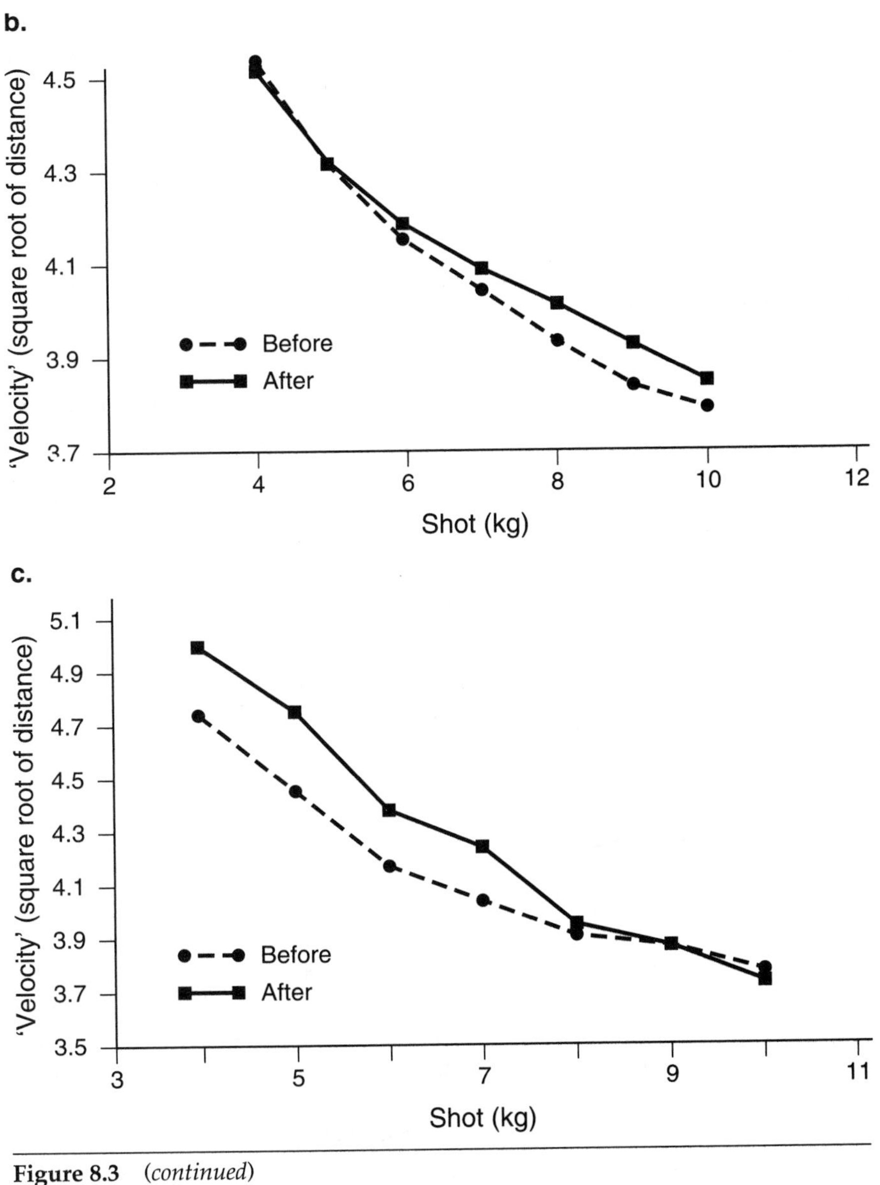

Figure 8.3 *(continued)*

■ *Training Goal: Muscle Power*

Perform the main sport exercise with added resistance. This often is the quickest way to make gains in athletic performance. It is also insufficient. The performance results initially advance but soon stop improving due to accommodation. Other training means are then necessary.

Enhance maximal strength. It is impossible for athletes to generate a large force in a fast movement if they cannot develop similar or even greater force values in a slow motion. But don't overemphasize the role of maximal strength in power production. To be a strong athlete does not mean to be a power athlete. It is true that all elite power athletes are very strong people. On the other hand, not all strong individuals can execute movement powerfully when combining large force and high velocity.

Train rate of force development (RFD). If the time available for force development is short, RFD is more important than maximal strength. Enhance not only maximal (F_{mm}) but also dynamic strength—the force developed at a high velocity of movement. Utilize drills requiring the utmost muscular effort against moderate resistance (the method of dynamic effort). Employ specific drills and methods to improve reversible (stretch-shortening) muscle action. This is a specific motor ability.

Muscle Mass

While muscle hypertrophy is a primary goal of body builders, athletes do not typically aim at increasing muscle mass. However, muscle hypertrophy is an important way to increase muscle strength. Furthermore, some athletes (linemen in football, throwers) are able to use heavy body weight to their advantage and thus want to increase muscle mass. Accordingly, methods that are essentially bodybuilder techniques are used by competitive athletes, too.

The main objective of such a training routine is the maximal activation of protein catabolism (breakdown of muscle proteins), which in turn stimulates the synthesis of contractile proteins during rest periods. Since the total amount of degraded protein is maximal when loads ranging between 5 to 7 and 10 to 12 RM are lifted, this specific training intensity (repeated effort and submaximal effort methods) is recommended. Training protocols are designed with the same primary objective, to activate the breakdown of proteins in the chosen muscle groups. In particular:

- Rest intervals between sets are short—1 to 2 min compared to 3 to 5 min in weight lifting training when the aim is to emphasize neuronal output.

- In one workout or even in one day, no more than two to three muscle groups or body parts are exercised. Then, on the following day, exercises for other muscle groups are included. This is called *split training*. An example is given in Table 8.1. With the split system, a muscle group is fully exhausted during a workout and then given time to recover (in this example, about 72

Table 8.1 An Example of the Split System of Heavy Resistance Training

Day 1	Day 2	Day 3	Day 4
Arms and shoulders	Legs	Chest	Rest
Abdominal muscles		Back	

hr). The muscle group is exercised twice a week. The split system is never used for perfecting the neural mechanisms of strength enhancement.

- Several exercises (usually from two to five) for the same muscle group are employed during a single training unit. Exercises may vary within the sequence; for instance, a curl with a dumbbell can be performed with the hand alternately in the supinated and pronated position. However, this is not done to alternate the muscle groups; that is, initially all exercises of one muscle group should be executed. For instance, all back exercises are performed first, and then chest exercises are performed. The idea is the same, to activate and exhaust the muscle group as much as possible. Exercises for the same muscle group, slightly changed from each other, are performed consecutively. This method, called *flushing*, was initially based on the assumption that increased blood circulation stimulates muscle growth. Up to 20 to 25 sets per muscle group may be executed in one workout. Table 8.2 summarizes the comparison between training to emphasize muscle mass and training to emphasize strength.

■ *Training Goal: Muscle Mass*

Activate the breakdown of proteins in the chosen muscle groups during training workouts and protein supercompensation during rest periods. Use weights with RM between 5 to 6 and 10 to 12 (the repeated effort and submaximal effort methods).

Follow the recommendations given in Table 8.2.

Endurance Performance

Endurance is defined as the ability to bear fatigue. Human activity is varied, and the character and mechanism of fatigue are different in every instance. Fatigue caused by work with a finger ergograph, for instance, has little in common with the fatigue of a marathon runner or a boxer. Thus, the corresponding types of endurance will differ.

Table 8.2 Training Protocols to Induce Muscle Hypertrophy or Muscle Strength (Neural Factors)

Training variable	Muscle hypertrophy	Strength (neural factors)
Intent	To activate and exhaust working muscles	To recruit the maximal number of motor units with optimal discharge frequency
Intensity, RM	From 5–7 to 10–12	1–5
Rest intervals		
- Between sets	Short (1–2 min)	Long (3–5 min)
- Between workouts emphasizing same muscle groups	Long (48–72 hr)	Short (24–48 hr)
Exercises in a workout	Three or fewer muscle groups (split system)	Many muscle groups
Exercise alternation in a workout	Flushing: exercises for the same muscle group may alternate; exercises for various groups do not alternate	Recommended
Training volume: load • repetitions • sets	Larger (4–5 times)	Smaller (4–5 times)

Muscular Endurance

Endurance of muscles is manifested in exercises with heavy resistance, such as the repetitive bench press, that do not require great activation of the cardiovascular and respiratory systems. Fatigue is caused by the functioning of elements in the neuromuscular system that are directly involved in the execution of the movement.

Muscular endurance is typically characterized either by the number of exercise repetitions one can carry out until failure (the maximum number of pulls up, squatting on one leg), or by the time one can maintain a prescribed pace of lifts or a posture. In either case, the load can be set in terms of absolute values such as kilograms or newtons (e.g., a 50-kg barbell) or in relation to the maximal force (e.g., a barbell 50% of F_{mm}). Accordingly, the *absolute* and *relative* indices of endurance are determined. In estimating the absolute endurance, individual differences in muscular strength are ignored. Everyone is asked to press the same weight, for instance. When relative endurance is measured, on the other hand, all are asked to press a weight that equals the same percentage of their maximum strength.

The absolute indices of endurance show considerable correlation with muscular strength; individuals of great strength can repeat a vigorous exercise more times than those of lesser strength (Figure 8.4). However, this correlation is observed only with resistance that is at least 25% of maximum strength. When the load is smaller, the number of possible repetitions quickly rises and is, in practical terms, independent of maximal strength (Figure 8.5). Relative indices of muscular endurance do not correlate positively with maximal strength. In fact, they often show negative correlations.

Let's consider an example of what we have seen about the correlation between strength and endurance. Suppose two athletes can bench press weights of 100 and 60 kg, respectively. It is obvious that the first athlete can press a 50-kg weight more times than the second athlete and that the absolute indices of endurance for the first athlete will be better. If both

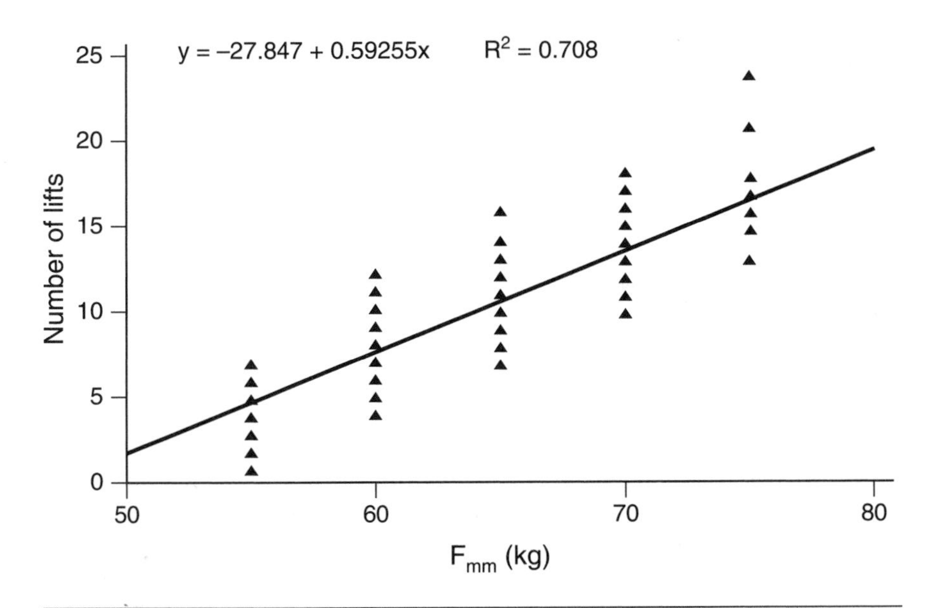

Figure 8.4 Maximal weight lifted in the bench press (F_{mm}, kg) versus the number of lifts of a 50-kg barbell in the same movement. The pace of lifts was 1 lift every 2 s. The subjects were wrestlers 16 to 18 years old ($n = 60$). The average value of the maximal strength was 65.7 kg. So the weight lifted (50 kg) was equal to approximately 75% of the average F_{mm} of the sample. The number of experimental points in the graph (41) is less than the number of subjects (60), since performance of some athletes was identical. When F_{mm} and the number of lifts were the same, several points coincided. Note. The data are from "Two Types of Endurance Indices" by V.M. Zatsiorsky, N. Volkov, and N. Kulik, 1965, *Theory and Practice of Physical Culture*, **27**(2), pp. 35-41. Reprinted by permission from the journal.

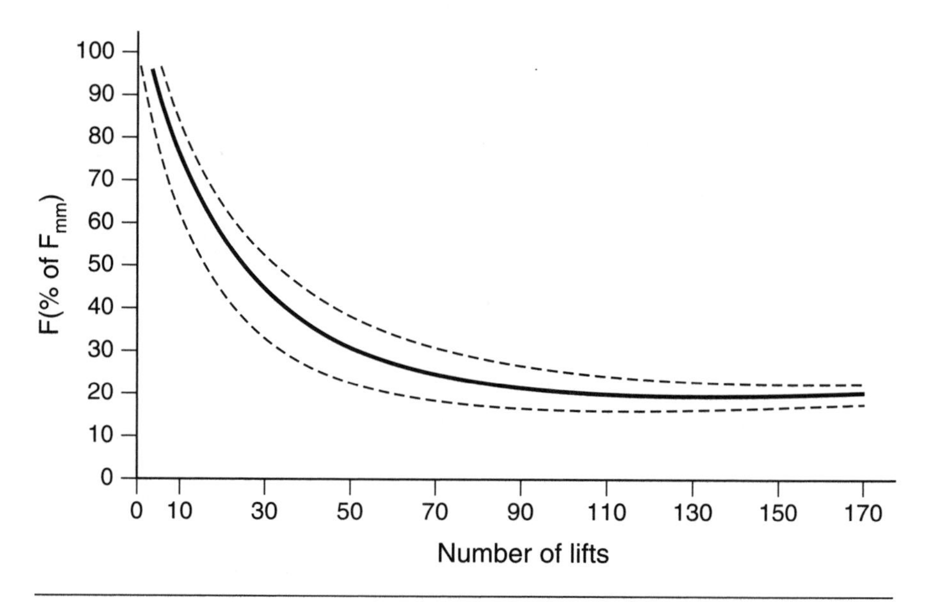

Figure 8.5 The dependence of the number of bench press lifts on the relative weight of a barbell. Average data of 16 weight lifters; the solid line represents rounded average figures; the broken line is for the standard deviation. Note. Adapted from "Two Types of Endurance Indices" by V.M. Zatsiorsky, N. Volkov, and N. Kulik, 1965, *Theory and Practice of Physical Culture*, **27**(2), pp. 35-41. Adapted by permission from the journal.

athletes are told to press a weight of 10 kg (less than 25% of maximal strength for each), it is impossible to predict who will exhibit more endurance. In this case endurance (measured by the number of repetitions) does not depend on strength level. If both athletes press a weight equal to 50% of their maximal force production (50 and 30 kg, respectively), it is again impossible to predict who will show greater endurance. Here, too, endurance does not correlate with strength.

Since athletes are not matched during a competition according to strength, practice should focus on absolute endurance. As we have noted, these indices are essentially dependent on the strength level; as the resistance an athlete must overcome increases, so does the dependence. Thus, when it is necessary to repeatedly overcome considerable resistance (more than 75-80% of the maximum muscular strength), there is no need for special endurance training. When resistance is smaller, though, one must concentrate on the development of both strength and endurance. In gymnastics, for instance, an athlete who cannot hold a cross for 3 s during a ring exercise (as the rule requires), still must train strength, not endurance. But a gymnast who performs four crosses in one combination and cannot hold a fifth must train endurance (together with strength).

Repeatedly performing strength exercises with resistance comprising 40% to 80% of maximum strength is the recommendation in this case. The repetitions are performed as many times as possible. If the magnitude of resistance is less than 20% to 25% of the athlete's strength, strength training (i.e., a training routine directed at increasing maximum strength) does not immediately improve athletic performance. Athletes from these sports, such as marathon runners, rarely use heavy resistance training.

To estimate the potential merit of strength training in a given sport, we should compare the force developed by an athlete during the main sport exercise to the individual's maximum strength during a similar motion. For instance, in a single scull, elite rowers apply an instantaneous force of up to 1,000 N to the oar handle. In dryland conditions, they generate forces of 2,200 to 2,500 N in the same posture. This means that during rowing, the athletes must overcome a resistance equalling 40% to 50% of their F_{mm}. Since the proportion of the force generated during the main sport movement is high, there is no doubt that strength training directed toward enhancement of maximum strength is useful for the rowers. However, it should be combined with muscular endurance conditioning.

Circuit training is an effective and convenient way to build muscular endurance. Here a group of trainees is divided into several (7-12) subgroups according to the number of stations available. Each trainee performs one exercise at each apparatus (station) as though completing a circle (Figure 8.6). Body-weight-bearing exercises, free weights, and exercise machines as well as stretching exercises may be used at different stations. Consecutive stations should not consist of exercises involving the same muscle groups. Trainees move quickly from one station to the next with a short rest interval in between each. The circuit is finished once the exercises at all stations are completed. The time for one circuit is prescribed.

All the characteristics of training programs (specificity, direction, complexity, and training load) can be easily specified and modified within a general framework of circuit training. However, in practice, only a limited variety of circuit programs are in use. Typically, circuit training routines use resistance of 50% to 70% of 1 RM; 5 to 15 repetitions per station; interstation rest intervals of 15 to 30 s; one to three circuits; and a total duration of 15 to 30 min.

■ *Training Goal: Muscular Endurance*

Compare the magnitude of force (F) generated in the movement of interest (for instance, during each stroke in rowing) with the maximal force values (F_{mm}) attained in the same motion during a single maximal effort in the most favored body position.

If $F > 80\%$ of F_{mm}, don't train endurance. Train maximal strength. If $F < 20\%$ of F_{mm}, don't train maximal strength. Train endurance. If

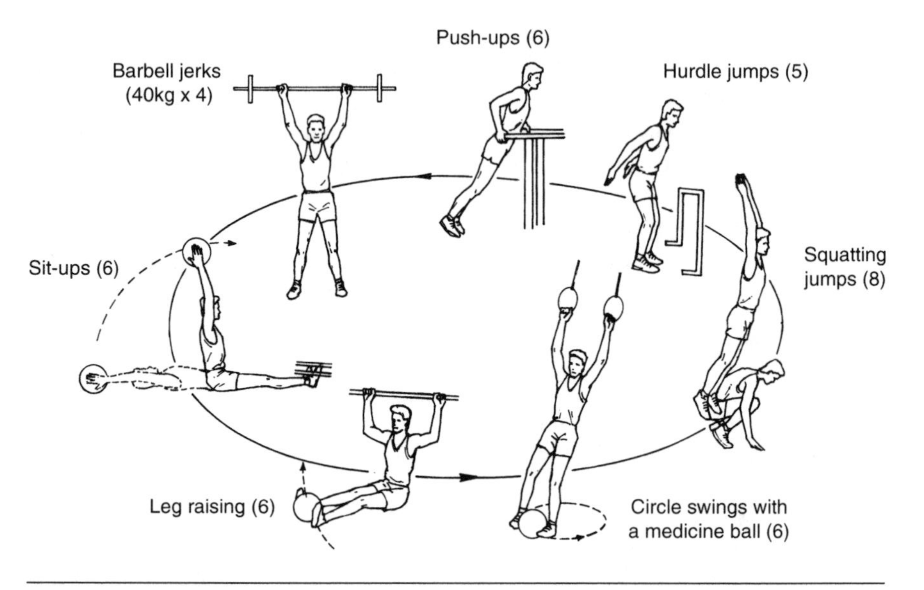

Figure 8.6 An example of circuit training. Note. From *Motor Abilities of Athletes* (p. 156) by V.M. Zatsiorsky, 1966, Moscow: FiS. Reprinted by permission from Fizkultura i Sport.

$20\% < F < 80\%$ of F_{mm}, train both maximal strength and muscular endurance. Utilize the method of submaximal effort. Vary the magnitude of resistance. Exercises in a set must be performed until failure. Employ circuit training.

Endurance Sports

In endurance sports, high energy demands are met by increased oxygen consumption as well as augmented anaerobic metabolism. The cardiovascular and respiratory systems become highly active. Athletic performance is limited by the central systems of circulation, respiration, and heat dissipation rather than peripheral muscle function alone. The correlation between local and general endurance has been shown to be small. Attempts to limit the off-season training of endurance athletes (speed skaters and skiers) to local endurance exercises (one-leg squatting) proved unsuccessful. Trainees improved their performance in one-leg squatting from 30 to 50 times to several hundred times (and even to more than 1,000 times) without any substantial improvement in the main sport. Because of these findings, strength training on the local involvement of a given percentage of the body's musculature was not popular among endurance athletes for many years. It was considered a waste of time and effort.

This is not the case, however, in contemporary sport. Since improvements in both endurance and strength are desirable for optimum performance in many sports, strength exercises are now extensively used by endurance athletes. However, the intent is not to enhance maximal strength per se, but (and this is the most important part of the concept) to enhance the force generated by the slow motor fibers. Recall that human muscles are composed of different fiber types, roughly classified as slow and fast. Slow motor fibers are highly adapted to lengthy aerobic muscular work. Fast motor fibers, adapted to short bursts of muscle activity, are characterized by large force and power output and high rates of force development. In the main, strength training is directed at increasing maximal muscular force production and thus primarily addresses maximal involvement of the fast motor units (MUs) and their strength gain. However, early involvement of the fast MUs in endurance activities vitalizes anaerobic metabolism and elicits early fatigue.

In endurance sports, the objective is exactly the opposite. Here the athlete wants to work as long as possible at a given intensity involving the slow motor fibers. In this case only, the metabolic response to exercise is aerobic and the athlete's work is sustained. The recruitment of fast motor fibers during prolonged work is apparently not desirable. The less the proportion of the activated fast motor fibers the better. So the force repeatedly exerted by an athlete during an endurance exercise should be compared not with maximum strength but with the maximum sustainable force by the slow (fatigue resistant, oxidative) motor fibers alone.

The slow motor fibers do not adapt to the enhanced force demand with "classical" methods of strength training. These methods are chiefly designed to recruit and train the fast motor fibers. Relatively low resistance and long exercise bouts are used to enhance the strength potential of endurance athletes. The corridor of MUs subjected to a training stimulus should presumably include slow MUs. Among coaches, it is a common belief that muscles must work at the highest levels of their aerobic capacity. Exercise sets comprising, for instance, 5 min of repetitive lifts are common. While training in the 1980s, world record holder and several-time Olympic champion in 1500-m swimming, Vladimir Salnikov, performed up to 10 exercise bouts on a special exerciser during dryland training. Each set was 10 min long. This routine, classified by swimming experts as strength training, only slightly resembles the training protocols used by weight lifters.

Such strength training is even more difficult to combine in training programs with endurance types of activity. The demands of the two types of activity are different. Heavy resistance training, for example, stimulates muscle fiber hypertrophy, which reduces capillary density and mitochondrial volume in the working muscles. These changes are detrimental to endurance. Endurance training, in contrast, elicits an increase in capillary density and mitochondrial volume density and may cause a decrease in muscle fiber size. When strength and endurance training are done concurrently, it is difficult for an organism to adapt simultaneously to the conflicting demands. Consequently, the combination of endurance and strength training impairs strength gains in comparison to strength training alone. This is also true with respect to endurance training. As the time between the two types of exercises lessens, the impediment becomes greater. Same-day training, for instance, impedes development to a greater extent than does training on alternate days. Another factor that influences the interference is the magnitude of the training load; the greater the load, the more incompatible strength training is with endurance training.

The solution is to conduct sequential strength and endurance programs, focusing first on strength training and afterwards on endurance (Figure 8.7). It is less efficient to proceed in the other order.

The motor ability that is not the prime target of training during a given mesocycle should be maintained with a retaining training load (except during the tapering period, when a detraining load is appropriate).

■ *Training Goal: General (Cardio-Respiratory, Especially Aerobic) Endurance*

Try to enhance the strength of slow motor units (fibers) that are oxidative and fatigue resistant. Don't use maximal weight loads. Utilize submaximal weight loads in combination with a large number of repetitions. Apply strength and endurance programs sequentially.

Injury Prevention

Heavy resistance training results in both increased muscular strength and increased mechanical strength of connective tissue structures around a joint (tendons, ligaments, ligament-bone junction strength). Strength training increases bone mineral content. A stronger muscle absorbs more energy than a weak muscle before reaching the point of muscle injury. This may be important for injury prevention.

To plan training routines to reduce the risk of injury, it is necessary to consider (a) muscle groups and joint motion, (b) muscle balance, and (c) coordination pattern.

Muscle groups that need to be strengthened can be classified as nonspecific and specific (actively involved in a given sport). The most important nonspecific muscle groups, which should be intentionally trained by young athletes regardless of the sport, are the abdominal muscles and

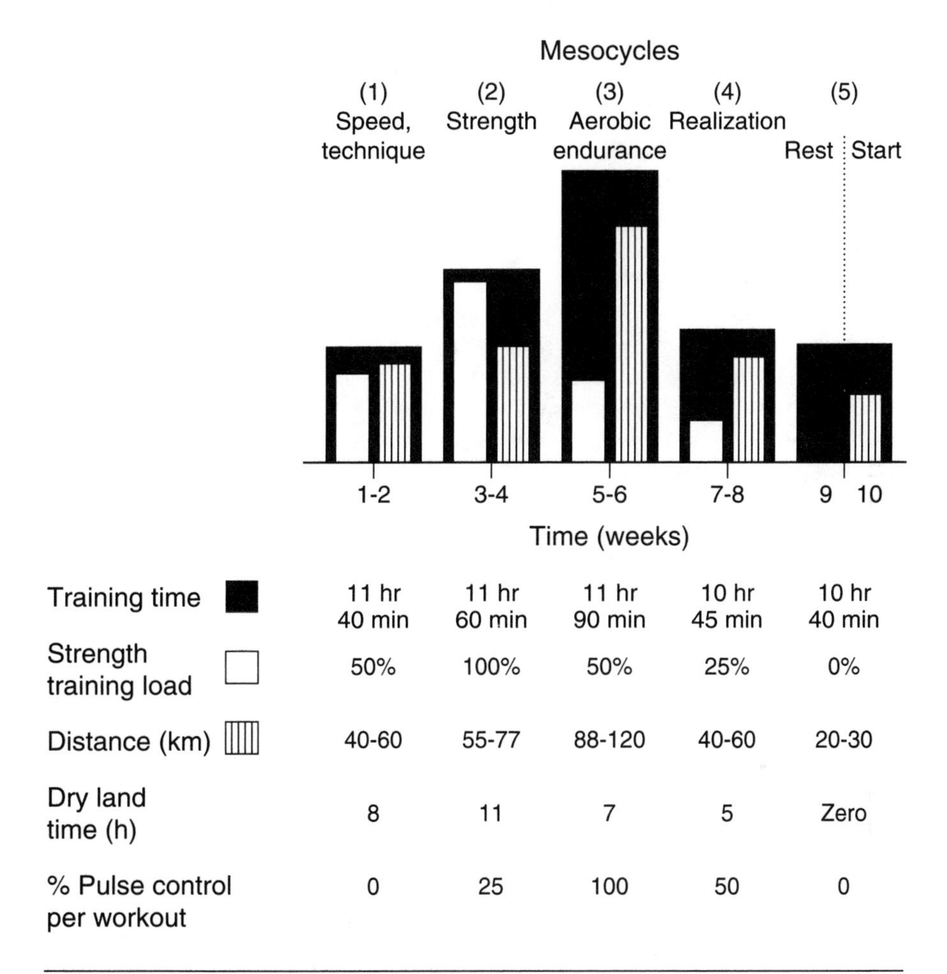

	(1)	(2)	(3)	(4)	(5)
Training time ■	11 hr 40 min	11 hr 60 min	11 hr 90 min	10 hr 45 min	10 hr 40 min
Strength training load □	50%	100%	50%	25%	0%
Distance (km) ▥	40-60	55-77	88-120	40-60	20-30
Dry land time (h)	8	11	7	5	Zero
% Pulse control per workout	0	25	100	50	0

Figure 8.7 Training plan of Vladimir Salnikov (1980 and 1988 Olympic champion in 1500-m swimming) developed by his coach, Igor Koshkin. Note the following: (1) The strength training mesocycle (2) precedes the endurance-oriented training. (2) Nontargeted motor ability is maintained with a retaining training load. The retaining load is roughly two to three times less than the stimulating one. For instance, dryland training time per week was 11 hr during the strength-oriented mesocycle and only 5 hr during the realization mesocycle; swimming distance per week was maximally 120 km and minimally 40 km. (3) Nontraditional short mesocycles, only 2 weeks long; usually 4-week mesocycles are used. Note. The data are from *Preparation of National Swimming Team to 1980 Moscow Olympic Games*, 1981, technical report #81-5 (p. 241), Moscow: All-Union Research Institute of Physical Culture.

trunk extensors. Improving the strength of these muscle groups is desirable to develop a base for intensive training. Specific muscle groups are different in diverse sports and may vary from the neck muscles (football, wrestling) to the small foot muscles (jumping, sprinting).

Muscles and joint structures need to be strengthened not only for joint movements that take place in the main sport exercise but also for other angular joint movements. It is especially important to strengthen joint structures in lateral movements (abduction-adduction) and in rotation relative to the longitudinal axis of a body segment (foot eversion-inversion, for example). For instance, football players usually perform many exercises to increase the strength of knee extensors. However, knee injuries are often caused by a lateral force acting during sideways movements or collisions. If the muscles and joint structures that resist lateral knee movement are not strengthened, the injury risk is very high. The same holds true for ankle motion. If only plantar flexion is trained, the athlete cannot resist high lateral forces acting on the foot. The strength may be too low to prevent hyperinversion (or hypereversion) and consequently trauma. Unfortunately, strength exercise machines, which are so popular now, provide resistance in only one direction—they have only one degree of freedom. Thus, the user does not have to stabilize the working parts of the apparatus as in exercising with free weights. Athletes accustomed to exercise machines lose a very important facet of motor coordination—joint stabilization. Even when the aim of strength training is to increase joint stability (for instance, in the case of knee laxity after a trauma), many athletic trainers and physical therapy specialists recommend exercising with isokinetic apparatuses, that permit knee flexion and extension only. Lateral movements, however, are not trained; unfortunately, it is exactly these muscles and structures that should be the training target.

Muscle balance is also important to prevent injury. First, a large imbalance in strength between the two legs should be corrected. If one leg is substantially stronger than the other, the running athlete performs a more powerful takeoff with the stronger leg and then lands on the contralateral weak leg, which is then systematically overloaded and at greater risk of injury. A difference of 10% or more in the strength of the two legs, or a difference of more than 3 cm in thigh circumference, necessitates exercising the weak leg. A second type of imbalance that should be avoided is between muscles and their antagonists (for instance, quadriceps and hamstrings). The force for knee extension is generated by the quadriceps, while deceleration of the tibia is the function of the hamstrings, which absorb the energy provided by the quadriceps. When the muscles are imbalanced such that the quadriceps are relatively stronger, hamstring overloading can result. Researchers have suggested that, to minimize the risk of injury, hamstring strength must be not less than 60% of quadricep strength. This recommendation is valid for strength values measured at the joint angular velocity 30°/s.

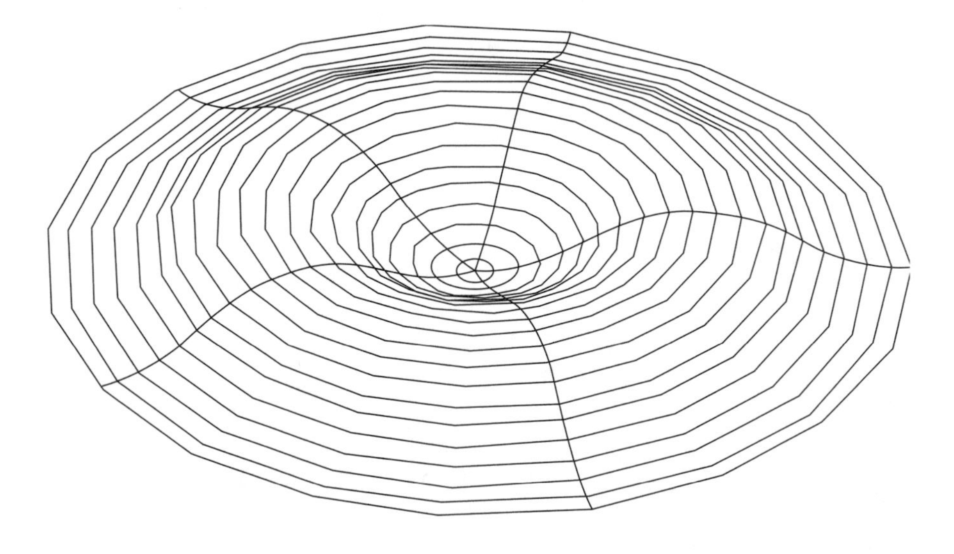

Figure 8.8 A "wave" surface for jumping exercises. Repetitive jumps with different foot placements (eversion, inversion, dorsiflexion) are performed on this device made of solid plastics. Maximal force is developed when the range of motion in ankle joints is near maximum, thus permitting the athlete to adapt to situations where an injury is typically incurred. Both strength and mobility in ankle joints can be improved. U.S. patent pending.

Finally, pay attention to the *coordination pattern of strength exercises*. The majority of movements include the stretch-shortening cycle, and injuries often occur during stretching phases of these cycles or during the transition from stretching to shortening, when muscle force is maximal (see chapter 2). Thus, exercises designed to decrease susceptibility to trauma should include reversible muscle action. Properly scheduled exercises employing reversible muscle action are useful for preventing injury. In these exercises, muscles are trained in natural conditions. Proper coordination patterns, muscle strength, and flexibility are all improved at the same time. As an example, repeated jumps on a specially designed "wave" surface (Figure 8.8), when performed regularly and correctly, strengthen the anatomical structures of the ankle joint and reduce ankle sprains and dislocations.

■ *Training Goal: Injury Prevention*

Strengthen the abdominal muscles and trunk extensors. Strengthen muscle groups specific to your sport. Muscles must be strengthened for both the joint movements that take place in the main sport exercise and for other angular joint movements. Correct imbalance of antagonists as well as disparities in muscular strength between the extremities. Employ drills encompassing reversible muscle action.

Summary

The general idea in *strength training for strength performance* is not to train strength itself as a unified whole; rather, it is to train the underlying muscular and neural factors. To improve neuromuscular coordination (MU recruitment, rate coding, MU synchronization, entire coordination pattern), the best method is maximal effort. On the other hand, to stimulate muscle hypertrophy, repeated and submaximal effort methods are more appropriate.

Although maximal strength is regarded as a prerequisite for high movement speed, its transmutation into velocity, or power, requires main sport exercises with additional resistance and assistance exercises to develop maximal strength, the rate of force development, dynamic strength, and force produced by stretch-shortening (reversible) muscle action. If there is not enough time for development of maximal force values in a sport exercise, the rate of force development (RFD) rather than maximum strength must be the primary training objective.

The aim of exercises designed to increase muscle mass is to break down proteins in particular muscle groups; this in turn stimulates the synthesis of contractile proteins during rest periods. The most effective loads for this type of training range between 5 to 7 RM and 10 to 12 RM.

Endurance is defined as the ability to bear fatigue. Human activity is varied, and as the character and mechanism of fatigue are different in every instance, so is endurance. Muscular endurance is typically characterized by either the number of possible exercise repetitions until failure, or by the time one can maintain a prescribed pace of lifts or a posture. In either case, the load can be set with absolute values (e.g., lifting a 50-kg barbell), or relative to the maximal force (lifting a barbell 50% of F_{mm}). With resistance greater than 25% of maximum strength, the absolute indices of endurance correlate positively with muscular strength. Relative indices of muscular endurance often correlate negatively with maximal strength.

Since athletes are not matched during a competition according to strength, practice should focus on absolute endurance. Athletes for whom resistance typically is less than 75% to 80% of maximal muscular strength should concentrate on the development of both strength and endurance. To estimate the potential value of strength training in a given sport, compare the force developed by an athlete during the main sport exercise to the individual's maximum strength during a similar motion. Note that circuit training is an effective and practical way to build muscular endurance.

Whereas strength training for the most part aims at maximal involvement and strength development of the fast MUs, the objective in endurance sports is exactly the opposite—to work as long as possible at a given intensity while involving the slow motor fibers. "Classical" methods of strength training are not designed to train these slow fibers. Relatively low resistance and long exercise bouts are used to increase the strength potential of endurance athletes. The intent is not to enhance maximal strength per se, but rather to increase the force generated by the slow motor fibers.

Strength training is difficult to combine with endurance types of activity. When strength and endurance training are done concurrently, it is difficult for an organism to adapt simultaneously to the conflicting demands. The solution is to conduct sequential strength and endurance programs. Focus first on strength training and afterwards on endurance.

Training routines designed to decrease injury risk will address muscle groups and joint motion, muscle balance, and coordination patterns.

Plan training to strength both nonspecific and specific muscle groups (those actively involved in a given sport). The most important nonspecific muscle groups, which should be intentionally trained by young athletes regardless of the sport, are the abdominal muscles and trunk extensors. Muscles and joint structures also need to be strengthened not only for the joint movements of the main sport exercise but also for other angular joint movements. It is especially important to strength joint structures in lateral movements and in rotation relative to the longitudinal axis of a body segment.

Another aspect of injury prevention is avoiding or correcting imbalance of muscles and antagonists as well as imbalance in strength between the extremities. Finally, exercises designed to decrease the susceptibility to trauma should include reversible muscle action.

Glossary

abdominal hernia—The protrusion of an internal organ or a part of an organ through the abdominal wall.

abduction—Movement of a limb away from the median plane of the body.

absolute indices of endurance—Endurance as determined by asking a number of subjects to overcome resistance of the same magnitude (e.g., to lift a 50-kg barbell).

accentuation—Increasing muscular strength primarily at the angular position at which maximal efforts are developed during the main sport movement.

accommodating resistance—Increasing muscular strength throughout the complete range of joint motion.

accommodation—Decrease in the response of a biological object to a continued stimulus.

accumulation mesocycle—Conducted to enhance the athlete's potential, that is, to improve the basic motor abilities (conditioning) as well as sport technique (motor learning).

actin—One of two proteins in muscle filament, the other being myosin.

acute (training) effects—The changes occurring during exercise.

adaptation—The adjustment of an organism to its environment.

adduction—Movement of a limb toward a median plane of the body.

adenosine triphosphate (ATP)—A biochemical substance used by all cells as an immediate source of energy.

aerobic—Utilizing oxygen.

aerobic exercise—Exercise during which energy is supplied by inspired oxygen.

afferent—Toward the central nervous system.

agonistic muscles—Muscles that initiate and carry out motion.

A-gradient—The ratio used to quantify the rate of force development in the late stages of explosive muscular effort; A-gradient = $F_{0.5}/(T_{max} - T_{0.5})$.

alpha-motoneuron—See **motoneuron**.

amino acids—Organic compounds (the "building blocks") that constitute muscle proteins.

amino acids, essential—Indispensable amino acids; must be provided by food.

anabolism—Synthesis of complex substances from simple ones; the opposite of catabolism.

anabolic—Promoting anabolism.

anaerobic—Without oxygen.

anatomical match—Coupling the inhalation phase with trunk extension and the exhalation phase with trunk flexion.

annulus fibrosus—A fibrous ring, the outer part of an intervertebral disk.

antagonist muscle—A muscle producing tension in opposition to the tension of another muscle.

ATP—See **adenosine triphosphate**.

axon —A nerve fiber.

ballistic stretching—A rapid stretching movement.

biomechanical match—Matching the expiration phase of breathing with the forced phase of movement, regardless of its direction or anatomical position.

calorie (cal)—A quantity of energy, especially heat.

catabolism—The disintegration of complex substances into simpler ones; the opposite of anabolism.

central factors (in force production)—The coordination of muscle activity by the central nervous system, including intramuscular and intermuscular coordination.

circuit training—Programs that consist of several "stations," with a specific exercise to be performed at each one.

clean and jerk—One of two lifts constituting the sport of weight lifting (Olympic style), in which the barbell is first lifted from the floor to the shoulders (clean phase) and then overhead (jerk phase).

competition period (of training)—On-season training.

compliance—The ratio of change in length per unit change in applied force.

concentric (or miometric) muscle action—Muscle shortening under tension, with the external resistance forces acting in the opposite direction from the motion.

contralateral—Pertaining to the opposite side of a body.

corridor (of motor units)—The subpopulation of motor units recruited and trained in a given exercise set.

cross-bridge attachment—The connection between the "head" of the myosin cross-bridge and the actin filament during muscle action.

cross-sectional area (of a muscle)—The area of muscle fibers on a plane perpendicular to their longitudinal axes.

cumulative (or accumulative) training effects—The result of the superimposition of many training sessions or even many seasons of training.

delayed muscle soreness—The pain and soreness that may occur 24 to 48 hr after training workouts.

delayed (training) effects—The changes manifested over a certain time interval after a performed training routine.

delayed transformation (of training load)—The delay of performance growth with respect to executed training work.

delayed transmutation (of nonspecific motor potential into sport performance results)—The time period needed to transform acquired motor potential into athletic performance.

detraining load—A load that leads to either a decrease in the performance results, the functional capabilities of an athlete, or both.

diaphragm—A musculomembranous wall separating the abdomen from the thoracic cavity.

diuretic—A drug that increases urine excretion.

doubled stress microcycle—Two stress microcycles in a row.

drag—The resistance to movement of a body offered by a medium, specifically air or water.

dynamic muscle action—Muscle lengthening or shortening under tension; see **concentric, eccentric**, and **reversible muscle action**.

eccentric (or pliometric) muscle action—Muscle lengthening under tension, with the external forces acting in the same direction as the motion.

efferent—Conducting impulses from the central nervous system.

efficacy coefficient (in periodization)—The proportion of athletes (%) who achieve their best performance during the most important competition of the season.

elastic—Resilient.

elasticity—The resistance provided by a deformed body, such as a rubber band or a spring.

electromyography (EMG)—Record of electric activity within or on the surface of a muscle.

endurance—The ability to bear fatigue.

energy—Capacity to perform work.

explosive strength—The ability to exert maximal forces in minimal time.

explosive strength deficit (ESD)—The relative difference between maximum maximorum force (F_{mm}) and maximal force (F_m) when the time available for force development is short; ESD (%) = $100 \cdot (F_{mm} - F_m)/F_{mm}$. ESD signifies the percentage of an athlete's strength potential not used in a given attempt.

extensor—A muscle that extends a limb or increases the joint angle.

external force—A force acting between an athlete's body and the environment; only external forces are regarded as a measure of an athlete's strength.

fascia—A fibrous membrane.

fast-twitch fibers—Muscle fibers that display high force, high rate of force development, and low endurance.

feedback—The return of output to the system.

fitness (physical fitness)—Slow-changing motor components of the athlete's preparedness.

fitness-fatigue theory—See **two-factor theory**.

flexor—A muscle that flexes a limb or decreases the joint angle.

force—An instantaneous measure of the interaction between two bodies, force being characterized by magnitude, direction, and point of application.

force feedback—See **Golgi tendon reflex**.

force gradient (S-gradient)—The ratio characterizing the rate of force development at the beginning phase of a muscular effort; S-gradient = $F_{0.5}/T_{0.5}$, where $F_{0.5}$ is one half of the maximal force F_m and $T_{0.5}$ is the time required to attain that force.

force-velocity relationship (curve)—The parametric relationship between maximal force and velocity values attained when a parameter of the motor task has been altered in a systematic way; motion velocity decreases as force increases.

generalized training theories—Simple models in which only the most essential features of sport training are taken into consideration.

Golgi tendon organ—A tension-sensing nerve ending located in series with muscle.

Golgi tendon reflex—The inhibition of muscle action evoked by a sharp rise of the pulling force applied to the muscle end.

human strength curve—See **strength curve**.

hydrodynamic resistance—The resistance provided by water.

hyperplasia—An increase in the number of cells.

hypertrophy—An increase in cell or organ size.

iliopsoas (muscle)—The compound iliacus and psoas magnus muscles.

immediate (training) effects—Effects that occur as the result of a single training session.

index of explosive strength (IES)—The ratio: IES = (The peak force)/ (The time to peak force).

indices of endurance—See **absolute indices of endurance; relative indices of endurance**.

individualization—Efforts to train according to the interests, abilities, and other particular characteristics of an individual.

inertia—Resistance due to the property of a body to remain at rest or to continue its movement in a straight line unless acted upon by an external force; a force is required to overcome inertia and to accelerate the body.

inertia wheel—A device used to study movement against inertial resistance in which the potential energy of the system is constant and all mechanical work, except small frictional losses, is converted into kinetic energy.

intensity coefficient (IC)—The ratio:

$$\text{IC}, \% = \frac{\text{average weight lifted}}{\substack{\text{athletic performance} \\ \text{(snatch plus clean and jerk)}}} \cdot 100$$

internal force—A force exerted by one constituent part of the human body on another part.

intervertebral disk—Disk of fibrocartilage located between two adjacent vertebrae.

intraabdominal pressure (IAP)—Pressure within the abdomen.

isokinetic—With constant speed; may refer to the rate of change of muscle length, velocity of the load being lifted, or angular velocity of the joint.

isokinetic muscle action—Muscle shortening at a constant rate; usually applied either to the constant angular velocity of a joint or to the constant linear velocity of a lifted load.

isometric (static)—Without change in muscle (or muscle plus tendon) length.

isometric muscle action—See **static muscle action**.

isotonic—With constant force; may refer to muscle action, a constant load being moved, or a constant joint torque over a range of motion.

length feedback—See **stretch reflex**.

load—Weight lifted; see also **training load**.

long-standing training (multiyear training)—Training embracing the entire career of the athlete, from beginning to end.

long-term planning (of training)—Planning multi-year training.

lumbar lordosis—Anterior convexity of the spine in the lumbar region.

macrocycle—One competition season; includes preparation, competition, and transition periods (phases).

maximal muscular performance—The best achievement in a given motor task when the magnitude of a motor task parameter (for instance, weight of an implement or running distance) is fixed; the symbol P_m (or V_m for maximal velocity, F_m for maximal force, etc.) is used throughout the book to specify maximal muscular performance.

maximal nonparametric relationships—See **nonparametric relationships**.

maximum maximorum performance (force, velocity, etc.)—Highest performances among the maximal, represented by the symbols P_{mm}, F_{mm}, V_{mm}; for instance V_{mm} and F_{mm} are the highest maximal velocity and force, respectively, that can be achieved under the most favorable conditions.

maximum training weight (TF_{mm})—The heaviest weight (one repetition maximum) an athlete can lift without substantial emotional stress.

maximum competition weight (CF_{mm})—The athletic performance attained during an official sport competition.

mechanical feedback—The impact of force generated either by an athlete upon the external resistance, by the movement performed, or both.

medium-term planning (of training)—Planning macrocycles.

mesocycle—A system of several microcycles.

microcycle—The grouping of several training days.

miometric muscle action—See **concentric muscle action**.

moment of force (or moment)—See **torque**.

motion—A movement determined only by its geometry; if all body parts move (in different attempts) along the same trajectory or very similar trajectories, the motion is considered the same, regardless of differences in force, time, velocity, and the like.

motoneuron (or motor neuron)—A nerve cell innervating muscle cells.

motor unit (MU)—A motoneuron and the muscle fibers it innervates.

muscle action—Development of muscle tension.

muscle fiber—A skeletal muscle cell.

muscle force arm—The shortest distance between the axis of joint rotation and the line of muscle action.

muscle spindle (stretch receptor)—A length-sensitive receptor located in muscle.

muscular corset—Muscles of the lumbar region.

muscular endurance—The type of endurance manifested in exercises with heavy resistance that do not require considerable activation of the cardiovascular and respiratory systems.

muscular strength—See **strength**.

muscle strength deficit (MSD)—The ratio: 100 · (force during electrostimulation–maximal voluntary force)/maximal voluntary force.

myofibril—A longitudinal unit of muscle fiber containing thick and thin contractile filaments.

myosin—Contractile protein in the thick filament of a myofibril.

myotatic reflex—See **stretch reflex**.

nonparametric relationship—The relationship between maximum maximorum performance (P_{mm}, V_{mm}, F_{mm}), on the one hand, and maximal performance (P_m, V_m, F_m, t_m), on the other; nonparametric relationships, in contrast to parametric ones, are typically positive.

neuron—A nerve cell.

nucleus pulposus—The cushioning, jellylike center within an intervertebral disk.

Olympic cycle (of training)—From one Olympic games to another; 4 years in length.

one-factor theory—A theory stating that the immediate training effect of a workout is a depletion of certain biochemical substances and that, after the restoration period, the level of the substance increases above the initial level (supercompensation).

overload—Training load (intensity, volume) exceeding a normal magnitude.

parameter—A variable, such as mass or distance, that determines the outcome of a motor task.

parametric relationship—Relationships between maximal force (F_m) and maximal velocity (V_m) attained in various attempts in the same motion (e.g., in shot-putting) when the values of the motor task parameter (e.g., shot mass) have been altered in a systematic way. The parametric relationship between F_m and V_m is typically negative: the greater the force (F_m), then the lower the velocity (V_m).

partial (training) effects—The changes produced by a single training exercise (e.g., bench press).

Pascal's law (of hydrostatic pressure)—A law stating that, in liquids, pressure is distributed equally on all sides; intradisk pressure measurements are based on this law.

peak-contraction principle—Increasing muscle strength primarily at the weakest ("sticking") point of a joint motion.

peaking—See **tapering**.

period of delayed transformation (of the training work into performance growth) —The time period between a peak training load and a peak performance.

period of training—System of several mesocycles.

periodization —A division of the training season into smaller and more manageable intervals (periods of training, mesocycles, and microcycles) with the ultimate goal of reaching the best performance results during the primary competition(s) of the season

peripheral factors (in force production)—The maximal force capabilities of individual muscles

pliometric muscle action—See **eccentric muscle action.**

power—Work per unit of time

preparation period (of training)—Off-season training

preparedness (an athlete's preparedness [AP])—The athlete's disposition for a competition, characterized by that person's potential sport performance

principle of progressive-resistance exercises—The progressive increase of resistance as strength gains are made.

puberty—Period in life at which sexual maturity is attained, occurring between the ages of 13 and 15 in boys and 10 and 16 in girls.

pubis—Pubic bone.

pyramid training (triangle pyramid programs)—Gradually changing the load in a series of sets in an ascending and then descending manner.

reactivity coefficient (RC)—$RC = F_{max}/(T_{max} \cdot W)$, where F_{max} is the peak force, T_{max} is the time to peak force, and W is an athlete's weight.

realization mesocycle (precompetitive mesocycle)—Planned to elicit the best sport performance within a given range of fitness

relative indices of endurance—Endurance as determined by asking subjects to overcome resistance that equals a specified percentage of their maximum strength (e.g., to lift a barbell 50% of F_{mm}).

repetition—The number of times a movement is repeated within a single exercise set.

repetition maximum (RM)—The maximal load that can be lifted a given number of repetitions in one set before fatigue; for instance, 3 RM is the weight that can be lifted in one set only three times.

residual (training) effects—The retention of changes following the cessation of training beyond time periods when possible adaptation takes place.

resistance (in strength training)—See **elasticity; inertia; weight; hydrodynamic; viscosity**.

rest interval—The time period between sets in a workout or between workouts.

retaining load—A load in the neutral zone at which the level of fitness is maintained.

reversible muscle action—Muscle action consisting of eccentric (stretch) and concentric (shortening) phases.

rule of 60%—The empirical rule that states the training volume of a day (microcycle) with minimal loading should be around 60% of the volume of a maximal day (microcycle) load.

sarcomere—The repeated contractile unit of a myofibril.

S-gradient—See **force gradient.**

short-term planning (of training)—Planning workouts, microcycles, and mesocycles.

snatch—One of two lifts constituting the sport of weight lifting (Olympic style), in which the barbell is lifted in one continuous motion from the floor to an overhead position.

specific exercises—Training drills relevant to demands of the event for which an athlete is being trained.

specificity—The similarity between adaptation gains induced by a training drill and the adaptation required by a main sport movement.

split table—An apparatus used for spinal traction.

split training—The training of different body parts on various days.

static (or isometric) muscle action—Muscle action at which muscle length is constant; no movement occurs.

sticking point—The weakest body position (joint angle) possessing minimal strength values.

stimulating load—A training load of a magnitude above the neutral level, eliciting positive adaptation.

strength (muscular strength)—The ability to overcome or counteract external resistance by muscular effort; also, the ability to generate maximum maximorum external force, F_{mm}.

strength curve (human strength curve)—The plot of force exerted by an athlete (or the moment of force) versus an appropriate body position measure (i.e., joint angle).

strength topography—The comparative strength of different muscle groups.

stress (impact) microcycles—Microcycles in which training loads are high and rest intervals are short and insufficient for restitution; fatigue is accumulated from day to day.

stretch receptor—See **muscle spindle.**

stretch reflex (myotatic reflex)—The contraction of a muscle in response to a stretch.

stretch-shortening cycle—See **reversible muscle action.**

supercompensation—An increase in biochemical substance content above the initial level after a restoration period following one or several workouts.

supercompensation phase—The time period in which there is an enhanced level of a biochemical substance after a workout.

superposition (of training effects)—Concurrent or sequential interaction of immediate and delayed partial training effects.

tapering (peaking)—The training phase occurring immediately before an important competition; combines features of the transmutation and realization mesocycles.

testosterone—Male sex hormone produced in the testes.

theory of supercompensation—See **one-factor theory.**

thick filament—A myofilament made of myosin.

thin filament—A myofilament made of actin.

three-year rule—The recommendation to use exercises with a heavy barbell (like barbell squats) only after 3 years of preliminary general preparation.

time deficit zone—A time period too short to generate maximum maximorum force.

timing of training—The distribution of exercises and training loads over certain time periods.

torque (moment of force) — The turning effect produced by a force; the external torque generated by a muscle is a product of the force generated by the muscle and the muscle force arm.

training effects—Changes that occur within the body as a result of training.

training load—An integral characteristic of the magnitude of performed training work.

training residuals—See **residual (training) effects.**

training session (workout)—A lesson comprising exercise and rest periods.

transfer of training results—The performance gain in a nontrained exercise.

transformation (of training work)—Performance gain as a result of adaptation to the executed training work.

transition period (of training)—The period of training immediately following a season.

transmutation (of motor potential)—The conversion of nonspecific motor potential into specific athletic performance.

transmutation mesocycle—Employed to transmute the increased nonspecific fitness into specific athlete preparedness.

triangle pyramid programs—See **pyramid training.**

two-factor theory—A theory according to which the immediate training effect after a workout is a superposition of two processes: gain in fitness prompted by the workout, and fatigue.

Valsalva maneuver—An expiratory effort with the glottis closed.

viscosity—The property of a semifluid, such as oil, that enables it to develop force dependent upon the velocity of flow.

weight—The resistance due to gravity.

work—Force times distance.

workout—See **training session.**

Suggested Reading

Atha, J. (1981). Strengthening muscle. *Exercise and Sport Science Reviews*, **9**, 1–73.

Baltzopoulos, V., & Brodie, D.A. (1989). Isokinetic dynamometry: Applications and limitations. *Sports Medicine*, **8**, 101–116.

Bell, G.E., & Wenger, H.A. (1992). Physiological adaptations to velocity–controlled resistance training. *Sports Medicine*, **13**, 234–244.

Bobbert, M.F. (1990). Drop jumping as a training method for jumping ability. *Sports Medicine*, **9**, 7–22.

Bompa, T. (1985). *Theory and methodology of training*. Dubuque, IA: Kendall/Hunt.

Buehrle, M. (Ed.) (1985). *Grundlagen des maximal—und Schnellkraft Training*. L. Schorndorf: Hoffmann.

Enoka, R.M. (1988). Muscle strength and its development. *Sports Medicine*, **6**, 146–168.

Fleck, S.J., & Kraemer, W.J. (1987). *Designing resistance training programs*. Champaign, IL: Human Kinetics.

Hopkins, W.G. (1991). Quantification of training in competition sports: Methods and applications. *Sports Medicine*, **12**, 161–183.

Jacobs, I. (1993). Adaptations to strength training. In D.A.D. Macleod, R.J. Maughan, C. Williams, C.R. Madeley, J.C.M. Sharp, & R.W. Nutton (Eds.), *Intermittent High Intensity Exercise* (pp. 27–32). London, E & FN Spon.

Kibler, W.B., Chandler, T.J., & Stracener, E.S. (1992). Musculoskeletal adaptations and injuries due to overtraining. *Exercise and Sport Sciences Reviews*, **20**, 99–126.

Knuttgen, H.G., & Kraemer, W.J. (1987). Terminology and measurement in exercise performance. *Journal of Applied Sport Science Research*, **1**, 1–10.

Komi, P.V. (1984). Physiological and biomechanical correlates of muscle function: Effects of muscle structure and stretch-shortening cycle on force and speed. *Exercise and Sport Science Reviews*, **12**, 81–121.

Komi, P.V. (Ed.) (1992). *Strength and power in sport: The encyclopedia of sports medicine*. Oxford: IOC Medical Commission, Blackwell Scientific.

Komi, P.V., & Håkkinen, K. (1988). Strength and power. In A. Dirix, H.G. Knuttgen, & K. Tittel (Eds.), *The encyclopedia of sports medicine: Vol. 1. The Olympic book of sport medicine* (pp. 181–193). Oxford: Blackwell Scientific.

Kraemer, W.J., Deschenes, M.R., & Fleck, S.J. (1988). Physiological adaptations to resistance exercise: Implications for athletic conditioning. *Sports Medicine*, **6**, 246–256.

Kulig, K. Andrews, J., & Hay, J.G. (1984). Human strength curves. *Exercise and Sports Science Reviews*, **12**, 417–466.

Matveev, L.P. (1981). *Fundamentals of sport training*. Moscow: Progress.

Mazur, L.J., Yetman, R.J., & Risser, W.L. (1993). Weight–training injuries: Common injuries and preventative methods. *Sports Medicine*, **16**, 57–63.

Plowman, S.A. (1992). Physical activity, physical fitness, and low back pain. *Exercise and Sport Sciences Reviews*, **20**, 221–242.

Sale, D., MacDougall, D. (1981). Specificity in strength training: A review for the coach and athlete. *Canadian Journal of Applied Sports Science*, **6**, 87–92.

Sekowitz, D.M. (1990). High frequency electrical stimulation in muscle strengthening: A review and discussion. *American Journal of Sports Medicine*, **17**, 101–111.

Stone, M., O'Bryan, H. (1987). *Weight training: A scientific approach*. Minneapolis: Bellwether Press.

Zatsiorsky, V.M., & Sazonov, V.P. (1985). Biomechanical foundation in the prevention of injuries to the spinal lumbar region during physical exercise training. *Theory and Practice of Physical Culture*, **7**, 33–40.

Index

About the Author

Vladimir M. Zatsiorsky, PhD, is a professor in the Department of Exercise and Sport Sciences at The Pennsylvania State University and a world-renowned expert in sport biomechanics and training athletes. Prior to coming to North America in 1990, Dr. Zatsiorsky served for 18 years as the department chair and professor of the Department of Biomechanics at the Central Institute of Physical Culture in Moscow.

For 26 years he served as consultant to the National Olympic Teams of the U.S.S.R. He was also director of the U.S.S.R.'s All-Union Research Institute of Physical Culture for 3 years, and was responsible for the scientific aspects of the preparation of all of the national teams for the 1988 Olympics in Seoul, South Korea.

In addition to his academic pursuits in the classroom, laboratory, and field, Dr. Zatsiorsky is a prolific writer who has authored or coauthored 230 scientific papers and more than 10 books. In recognition of his achievements, he has also received several awards, including the Geoffrey Dyson Award from the International Society of Biomechanics in Sport (the society's highest honor) and the U.S.S.R.'s national gold medals for the Best Scientific Research in Sport in 1976 and 1982.

Dr. Zatsiorsky is a member of the Medical Commission of the International Olympic Committee, International Society of Biomechanics, American Society of Biomechanics, and the Club of Cologne.